What is a designer: education and practice

To Mark, 1969, and those with him at Bristol, Hornsey, Guildford

What is a designer: education and practice

a guide for students and teachers

Norman Potter

Studio-Vista : London
Van Nostrand Reinhold Company : New York

A Studio Vista/Van Nostrand Reinhold Art Paperback

Edited by John Lewis

© 1969 Norman Potter

All rights reserved. No part of this work covered by the copyrights hereon may be reproduced or used in any form or by any means – graphic, electronic, or mechanical, including photocopying, recording, taping or information storage and retrieval systems – without written permission of the publisher.

Published in London by Studio Vista Limited
Blue Star House, Highgate Hill, London N19
and in New York by Van Nostrand Reinhold Company
450 West 33 Street, New York, NY10001
Library of Congress Catalog Card Number: 79-81761
Set in 9 on 12pt Times New Roman
Printed and bound in Great Britain by Tonbridge Printers Ltd
British SBN 289 79752 7 (paperback)
    289 79753 5 (hardbound)

Acknowledgments
My thanks to many students and colleagues for their part in this book; to Anthony Froshaug for detailed criticism; to John Lewis for urging it along. I also acknowledge helpful comment from the late Sir Herbert Read, Professor Misha Black, and Mr Grant Henke.

Contents

Introduction 7

PART I: Designers and their education 9
1 What is a designer? 9
2 Is a designer an artist? 14
3 Designer as craftsman 20
4 The nature of formal design education 23
5 What is good design? 38
6 Summary 47

PART II: Notes on design procedure 53
7 Introduction 54
8 When is an analytical approach necessary? 56
9 How is design work conducted? 64
10 Communication for designers 71
11 Drawings and models 76
12 Asking questions 78
13 Seeking information 82
14 Reports and report writing 83

APPENDICES
I Notes on some of the references 89
II Suggestions for beginners 92
III Extract from report on National Conference on Art and Design Education, London 96

Architecture is organisation. *You are an organiser, not a drawing board stylist.*
LE CORBUSIER

woe to the man whose heart has not learned while young to hope, to love – and to put its trust in life.
CONRAD

the disciplinary barriers are impenetrable. If these barriers in education were to vanish, the architect as benevolent dictator would vanish too. Instead students could arm themselves with useful tools and knowledge with which they could assist a community . . .
TOM WOOLLEY, President, British Architectural Students' Assn.

Introduction

A simple purpose of this book is implicit in its title: to ask what a designer does, how and why he does it, and what kind of education is open to him. This should help beginners and those who advise them.

I have also asked some less obvious questions that may concern students of all ages in a way that I hope will encourage all who want to do good work. I mean, of course, to sharpen their own tools of exploration; and to take heart.

This is a book of words about a realm of action; a designer thinking aloud from his own experiences. Such an approach cannot claim, or seek, the benefits of a disinterested critical scholarship. Nor should it be self-contained; the book is written to complement existing or alternative references – extending them in directions that are thinly considered elsewhere.

There are many well-illustrated books on design artefacts and this is not one of them. If students want to, I hope they will scribble diagrams all over the argument (and there are a few blank pages), which might be a designer's way of at once possessing an argument and departing from it. There are other ways, other designers, and other arguments.

There are three reasons for a book of this *kind*, as distinct from its achievement. First, to contribute some first-hand evidence toward a critical tradition that is still hardly waving its arms and making human noises. Second, to help students see more clearly what it is we have to confront; and thus, when it happens, to offer ourselves more directly and more generously. Third, we are in trouble. Perhaps there is nothing new in that. But there is a barrenness in our situation and I have tried to put my finger on some sense of it, while not forgetting what is more necessary; to tell upon one's pulses, as Rilke said, the sources of eloquent life.

NORMAN POTTER, April 1969

Part I: Designers and their education

1 What is a designer?

> Design:
> *v* to mark out, to plan, purpose, intend ...
> *n* a plan conceived in the mind, of something to be done ...
> *n* adaptation of means to ends ...
> *Shorter Oxford English Dictionary*

Every human being is a designer, which at first sight should commend this book to a sizeable readership. Many also earn their living by design – in every field that warrants pause, and careful consideration, between the conceiving of an action, a fashioning of the means to carry it out, and an estimation of its effects.

In fact this book is concerned with a professional minority: designers whose work carries a strong visual motivation, and whose decisions help give form and order to the amenities of life, in the usual context of manufacture. The clumsiness of this definition underlines the difficulty of using one word to denote a wide range of quite disparate experiences – both in the outcome of design decisions, and in the activity of designing. The dictionary reference above is selective; in practice the word is also applied to the *product* of 'a plan conceived in the mind', not only as a set of drawings or instructions, but as the ultimate outcome from manufacture.

This is confusing. The difficulty becomes acute if the word 'design' is used without reference to any specific context – used, for instance, as a blanket term to cover every situation in which an adaptation of means to ends is preceded by an abstract of intent – though designing is thus usefully distinguished from 'making' or from spontaneous activity. Beyond this point, the word must refer to recognizable products and opportunities, or become hopelessly abstract.

The design work to be discussed is generally taught in an art-school – the graphics part of it used to be called commercial art – but architects also are designers, drawing upon the resources of a cultivated visual imagination balanced by other and more technical areas of capability. Taking a broad view, it is convenient to group the work into three simple categories, though the distinctions are in no way absolute, nor are they always so described: product design (things), environmental design (places) and communications design (messages). Such categories blur some further necessary distinctions (as between, for instance, the design of industrial equipment and that of retail products in a domestic market) but can form a useful departure.

In the field of product design, the professional extremes might be said to range from studio pottery and textile design at one end of the spectrum, to engineering design and computer programming at the other. This is a very broad spectrum, and clearly there are serious differences at the extremes. In the communication field, a similar spectrum might range from, say, freehand book illustration, to the very exact disciplines of cartography or the design of instrumentation for aircraft.

Obviously, the more aesthetic and sensory latitude available within a particular range of design opportunities, the closer they resemble those offered by the practice of 'fine-art'. The less latitude, the closer design becomes to the sciences, and to fields in which the margin of aesthetic 'choice' is truly marginal. The design of a traffic light system has an aesthetic component, but it would need a very special definition of aesthetics to embrace the many determining factors that must finally settle the design outcome.

The situation for architects is usually held to be more straightforward; historically, their role has developed toward a fairly clear definition of responsibilities. However, the complex changes in building types, and in industrialized building situations, have combined with other factors thoroughly to upset this stable picture. Indeed, the architect's work has been so undermined by that of specialists in surrounding territory (engineers, planners, sociologists, interior designers etc.) that the profession is no longer so easy to identify. It is still reasonable to view an architect as a designer with a specialized technical and functional competence, and again a spectrum is discernible, ranging from very open and ephemeral design situations, to those as critical as the design of an operating theatre.

It is necessary to start somewhere, and this book takes a middle-zone standpoint. In most art schools this will include furniture, interior design, exhibition design, packaging, some wide areas of graphic and industrial (product) design, and some of the fringe territory leading into architecture. Students must make the necessary allowances to accommodate their own subject of study. This is chiefly necessary in the chapter 'Is a designer an artist?', and in some of the notes on procedure – or the studio potter will certainly feel that everything in this book is unduly complicated, whereas an architect might feel that there is undue simplification. All designers, however specialized, should know roughly what their colleagues do – and why; not only to fertilize their own thinking, but also to make group practice effective, and for other reasons that will be discussed.

There are many roles for different kinds of designer even within any given sector of professional work. A functional classification might divide designers thus: impresarios, culture diffusers, culture generators, assistants, and parasites. Impresarios: those who get work, organize others to do it, and present the outcome; culture diffusers: those who do competent work effectively over a broad field, usually from a stable background of dispersed interests; culture generators: obsessive characters who work in back rooms and produce ideas, often more useful to other designers than the public; assistants: often beginners, but also a large group concerned with administration or draughtsmanship; and parasites: those who skim off the surface of other people's work and make a good living by it. The first four groups are interdependent, necessary to each other. It should be added that any designer might shift from one role to another in the course of his working life, or even within the development of a single commission, though temperament and ability encourage a more permanent separation of functions in a large design office. Thus no value – judgement is implied here; except upon parasites who are only too numerous.

In small offices – or of course for independent free-lance workers – there will be little stratification; 'the office' may tend to move in one direction or another, but the work within it will be less predictable for any one member – excluding, perhaps, secretarial or administrative assistants and often temporary draughtsmen. A 'consultant' is often a lone wolf who deals in matters of high expertise or (paradoxically) of very broad generality. Designers will be found in every quarter; sometimes working independently, sometimes for government or local authority offices, or attached to large manufacturers, to retail agencies, to public corporations, and elsewhere in places too numerous to mention. There are a few design offices who will design anything from a fountain pen to an airport, and will therefore employ specialists from every field (including architects) – a rational development and a welcome one, but implying some genius for large scale organization which, in turn, may tend to level out the standard of work produced (as numbers increase, it becomes a problem to keep work flowing through at a productive pace, yet have enough – not too much – to allow everyone a fair living). Students usually need a few years' office practice before setting up by themselves; often this happens in small groups of three to six designers who will share office and administrative expenses.

Most designers are educated in a formal way by three-to-seven years in a design school (or school of architecture) leading to appropriate qualifications. Some have had unorthodox beginnings – by dropping in the deep end and learning to swim – but self-training may need sympathetic patrons, is apt to be patchy according to the opportunities that occur, and needs a special pertinacity. Apprenticeship rarely means more than training as a draughtsman. A few factories or retail firms may encourage employees who show design aptitude. Evening classes and correspondence courses are mostly directed at cultural-appreciation or do-it-yourself horizons, but intending full-time students can build up a portfolio of work by this means. Other kinds of training are mentioned in this book.

This, then, is the apparent situation of the designer and the starting point for this book. A note of warning: the word 'design' appears freely as noun and verb, and where words like 'formal', 'realization', 'consciousness' are used without qualification, readers should examine the context and think for themselves. I have used the word 'student' suggestively; trying it for size.

Returning to the statement that every human being is a designer, and using it as a springboard: we do well to remember that designers *are* ordinary human beings, as prone as others (given half a chance) to every human weakness, including an exaggerated idea of their own consequence. Consider the following questions: Should a designer design for a factory in which he could never imagine working as an operative? Is design social realist art? Is it handy to be in a state of moral grace when designing a knife and fork? Does design work justify its claims to social usefulness, or is it a privileged form of self-expression? Is a profession a genteel self-protection society with some necessary illusions? Should a designer be a conformist or an agent of change?

Students who feel that such questions are diversionary and a waste of time, should perhaps put this book down; others read on, but not for easy answers.

2 Is a designer an artist?

Before discussing this question, which involves describing a designer's work in some detail, it is necessary to look at the context in which it is usually asked. In England, it is certainly necessary to remember two things. First, the extraordinary cultural insularity of the last fifty years that permitted the early achievement of the English Arts & Crafts movement to be built into foundations of growth on the continent, while continuing a placid, homespun and largely arrested development in its place of origin. To illustrate this, it is only necessary to consider the lively interaction of media, disciplines and controversy in the de Stijl movement, and to take a sample of the English situation at the time, or to recall the strange scene in 1968 when so many students and teachers were to be seen confronting their first substantial awareness of the Bauhaus. The English book *Circle*, published in the late 1930's, marked a coming-together which, in English terms, was promising merely because unusual. An informal pointer to an absence of spirit can be seen from a comparison of the early edition of Herbert Read's *Art and Industry*, with layout by Herbert Bayer, and the subsequent 'tasteful' editions in which the spirit of the original – not in itself anything very remarkable – is absorbed back into the English literary traditions of book publishing. It was indeed in literature that things were more lively. The war destroyed even these tentative growths in a way that was more profound than is generally realized. The 1951 exhibition became a rallying point and proved to be a very odd mix of English empiricism – perhaps expressing itself in a renewed sense of place and occasion to set against a threatening Platonism – combined with belated public tribute to the 'international style' in building. In the matter of a robust critical climate, the author can testify to the difficulty of summoning a confrontation with the work at Ulm (Hochschule für Gestaltung) and its implications, even in 1960–62 in post-graduate art and design education.

In England, these factors must be taken into account. The second difficulty is more widespread, though stemming partly from comparable situations. It is the arbitrary split between what has to be called 'fine-art', namely painting and sculpture, and 'design' subjects, which all together happen to be taught in art-schools (frequently separated from architecture). It would be out of place here to examine the history of this problematic and to some extent (now) arbitrary grouping of studies. It is enough to point out that the situation could be more realistically appraised if painting and sculpture were studied alongside music, dance, poetry, film and other activities which interpret, primarily, the psychological and sensuous and spiritual understanding of man, as distinct from those activities which must satisfy first his physical and accessory needs under conditions of complex social constraint. In the last analysis, every human arte fact – whether painting, poem, chair, or rubbish bin – evokes and invokes the inescapable totality of a culture, and the hidden assumptions which condition cultural priorities. (In a basic sense, and given the conditions for warmth, food and shelter, the rest is a choice and speaks to us of priorities which need constant revaluation.) For the purpose of the remarks which follow, it is certainly necessary to say that if the words 'fine-art' and 'design' simply refer to a duality as experienced in art-schools, it is difficult to set up a satisfactory argument on that basis.

For the discussion that follows, the situation is seen from the standpoint of a designer.

Here is a sober but accurate description of professionalism by Professor Misha Black: '... the offering to the public of a specialized skill, depending largely upon judgement, in which both the experience and established knowledge are of equal weight, while the person possessing the skill is bound both by an ethical code and may be accountable at law for a proper degree of skill in exercising this judgement...'

Not, obviously, a full description, and perhaps a somewhat negative one, but making the fact plain that a designer works through and for other people, and is concerned primarily with their problems rather than his own. In this respect he can be regarded as rather like a doctor, with a responsibility for accurate diagnosis (problem analysis) and relevant prescription (design recommendations). It must be clearly realized that designers work and communicate indirectly, and their creative work finally takes the form of instructions to contractors, manufacturers and other executants. The exception is the designer-craftsman, whose position will be discussed later. The 'Instructions' may include written specifications, reports, and other documents, detailed working drawings, presentation drawings for clients, scale models and sometimes prototypes full-size. Since this is as far as a designer goes in actually making anything (strictly what he makes are visual analogues), it is vital that the instructions are both complete and fully intelligible to their recipients.

The designer usually has the further responsibility of supervising the work, but there is no obvious equivalent for the feedback through eye-and-hand so familiar to the painter or sculptor, whereby the original idea is constantly developed, enriched, or diverted by the actual experience of the materials and the making-process. For the designer the point of no return is indeed a moment of truth. Feedback does of course operate at the design stage, mainly through people, circumstance, and the continuous absorption of new information into the design brief. ('Brief' is a word for the designer's understood instructions from his client; his working terms of reference.) The outcome will still change radically from first ideas thrown up by superficial acquaintance with the design problem, but the changes will not always be of the designer's own choosing: their nature may be objectively determined by factors quite outside his control. Such factors might be something to do with costs, the availability of materials or techniques, a change in the client's requirements, or simply the discovery of factors that were hidden from sight in the early stages of the job.

Hence, in summary, the designer instructs, and his work involves many people, with some of whom he will have contractual relationships.

In some instances of design many specific responsibilities may arise – to clients, contractors, to the public who utilize the end-product, to numerous specialists or colleagues who may be involved if the undertaking is a large one (which implies team-work and frequently shared decision-making). If it is a building, it mustn't fall down; if it is a chair, it mustn't be thirty inches high, have an innate tendency to collapse under load, it mustn't employ joints that can't be made except by special machinery (unless this can be found economical) and it mustn't cost so much as to be unmarketable. The designer cannot exercise personal insights until every apparently conflicting factor in his brief has been reconciled to best advantage: until, in short, he knows exactly what he is up against and which constraints eventually – in fact – play in his favour.

For these reasons, the designer is highly 'problem' conscious; a large part of his work may consist in problem analysis, though rarely of the complex order familiar in the sciences. To an ability for sorting, ordering, and relating information he must bring qualities of judgement and discrimination as much as a subjective capacity to arrive at an imaginative solution. There is a diffuse sense in which the most seemingly 'objective' procedures in problem analysis are in practice discretionary, embedded as they are in the whole matrix of professional judgement in which relevant decisions are conceived. In some fields of design – e.g. textiles, there is far greater latitude than in others. In most design work the ultimate decisions affect, in a vital degree, appearance; but the look of the job, however crucial, emerges from its functional and circumstantial background.

Drawings can never be an end for a designer (excepting an illustrator); they are a means to the end of manufacture, and their expressive content is strictly limited to the purposes of relevant communication. This obvious distinction from fine-art drawing can easily be overlooked in a design school where the design projects are theoretical, and drawings become the only outcome, acquiring the false dignity of an end-product in the process. This does not imply that drawings can be loveless, slovenly, or inadequate in any way, but that their nature is strictly purposeful. It may indeed be necessary to the designer to make loving, scrupulous, over-adequate drawings for his own self-satisfaction and to preserve his own standards: but only in this sense are design drawings 'self-expression'.

At every stage of design there will be discussion, questions and argument; the final design will have to be demonstrated and if necessary defended to the client, who will not understand what the final result will look like, but will naturally tend to assume that he knows more about his own problems than does the designer – despite having called him in to solve them. A design solution intermingles with the world of considerations familiar to the client: communication media must be carefully chosen – verbal reports and other documents may accompany drawing and models. Designers use words constantly and in direct relation to their work; in forming and discussing ideas, assessing situations, annotating drawings, writing specifications and letters, and in report writing. This aspect of design work is frequently underestimated: an ability to use words clearly, pointedly, and persuasively is at all times relevant to design work.

It is now possible to ask, what kind of person might be happy and personally fulfilled in taking up design? It will be seen that a designer must be capable of more detachment than may be necessary to a fine-artist. He must be able to weigh up a problem dispassionately on its own terms (as well as his), and to select, arrange, and dispose his decisions accordingly. He must be able to thrive on constraint and to turn every opportunity to good account. He must like and understand people and be able to treat with them; he must be able to accept fairly complex situations in which he may well be working as a member of a team. He must be reasonably articulate. He must be practical and prepared for extensive responsibilities to other people. Finally he must be prepared to spend at least half his time working with graphic media, since most design work appears in drawings of one sort or another when decisions have been finalized.

These remarks may suggest an uncomfortably glum idea of human perfectibility. In practice, of course a designer's life is as muddled, informal and accident-prone, as most people's lives manage to be; not only behind the scenes, but sometimes in front of them. The point is that every profession has roughly defined public responsibilities, which are met as closely as possible by accepted codes of practice. Again, the fact that design work is ten per cent inspiration and ninety per cent fairly hard work – not an unusual prospect – does need some well-organized procedures to keep the brief clearly in view, and the available energies best occupied.

'Goal-seeking' is not unfamiliar to a fine-artist, and certainly as much hard work is involved. The work – in its origins, aims, and outcome – is far more subjective. A painter's first responsibility is to the truth of his own vision, even though that vision may (or maybe always does) change as his work proceeds. He may be involved with contractual responsibilities, but not to the same extent as is a designer, whose decisions will be crucially affected by them. The designer works with and for other people: ultimately this may be true of the fine-artist, but in the actual working procedure a designer's formative decisions have a different order of freedom. The fine-artist is less dependent on discussion, agreement, letters, reports, visits: the apparatus of communication that brings definition to a design problem, and relevance to its solution. A fine-artist usually works directly with his materials, or a very close visual analogue to the final work. As we have seen, the designer has a long way to go before decisions can emerge – and even then a model may be the closest 'tangible' embodiment of his ideas.

In the case of film, television, and theatre, which might be described as a realm of public art, quite complex design procedures are involved, similar in kind to many that were described earlier. In the main, however, the real connection between fine-artists and designers springs from the benefit of a shared visual sensibility; not from a relevant or direct transference of skills, language, or formative insight, from one field to the other. Students are warned that this is an opinion: recalling the breadth of the design 'spectrum', they will see that this is a difficult matter to unravel. So many factors impinge on the visual appearance of a design outcome that a designer's hand would seem to be guided by a wholly different 'requiredness' (a term borrowed from Gestalt psychology) from that which informs a painting or a work of sculpture. Yet there are component experiences with something in common. Equally valid transferences may occur from the 'feel' of related work in other fields (e.g. philosophy or mathematics). Similarly, a creative sensibility may derive from sources that cannot be looked for in any one field alone.

It is only necessary to hammer home the obvious because fine-art and design (excepting architecture) are often taught as closely inter-related subjects, and students are asked to choose between them. The term 'fine-art' is unpleasantly genteel, but will be met with in art school, generally to comprise painting and sculpture and to distinguish these studies from 'applied art', in the various fields of design discussed here. The view that there is a parallel situation in the sciences, as that of 'pure' science and 'applied' technology, is a questionable one: equally untrustworthy is the supposition that painting, sculpture, industrial design, architecture, derive in some sense from the common fountain-head of 'art'. To suggest this seriously requires a view of art (and a set of definitions) quite outside the scope of the present discussion: it is partly a semantic problem, pointing to the inadequacy of ordinary descriptive language. Without distorting common usage, it might be said that designers are content to bring a certain artistry to their work, and to recognize that there is much in common between the few masters in any field – fine-art, design, science, medicine, philosophy – more, perhaps, than unites the very disparate standards that co-exist in any one profession.

3 Designer as craftsman

Many designers have their own small workshop, for modelmaking, for making-up and testing full-size details, for prototype work, and sometimes simply for play – which is not only a relief from the drawing board, but can prove directly and unexpectedly helpful to any job on hand. The scale and equipment will vary a good deal, according to circumstance and the nature of the work. Industrial designers will usually have access to factory facilities for prototype work, or may cooperate with the research and development group already attached to the factory. In Scandinavia, it is not uncommon to have designers operating their own studio/workshop in direct association with the factory – a large proportion of their time being given to the production needs of the factory, but some to random experiment which may, or may not, ultimately prove valuable to their hosts. Product designers will generally design specifically for the production (and sales) potential of a given manufacturer, and will therefore need very close contact with factory management and machine operatives.

There are many fields of design in which conditions may be rather different. One-off jobs (e.g. shopfitting and exhibition stands) may be economically handled by smaller production units, and short production runs for special purposes may also be within their scope. Such work may include jobbing printing, exhibition and shopfitting work, purpose-made furniture for hostels, schools, laboratories, hotels, etc., interior conversion work including built-in fittings, ceramics, and other instances too numerous to mention. It is decidedly a mistake to assume that every serious design problem must refer back to the special conditions of mass-production, when in fact other opportunities are real enough and can often accommodate a more flexible design approach. The required production facilities will certainly include a high degree of mechanization, but the machines will be 'open-ended' in their input–output potential: that is, they will be of the kind that extend the hand rather than totally replace it. (E.g. a joinery shop of this sort would have a router, spindle, dovetailer, in addition to the usual sawbenches, planers, etc., but would be unlikely to have a multiple jig borer, variable offset lathes, or even a four-cutter.)

Obviously plenty of smaller firms meet these demands, and frequently a designer will be able to build up a useful working relationship with one of them. However, there is also scope for small units operated directly by the craftsman-designer or artisan, with a small number of employees or associates. (In the latter case, a co-operative can be a fruitful and a proper way of working, but is difficult to assemble and dispose all the required skills, if a unit is to be economically competitive and survive.) The potential can be an exciting one. The workshop experience is a curious amalgam of machine processes, creative accident, manual labour, and direct feedback from work in progress. It is an experience that may profoundly affect the attitudes of a designer who has worked in this way at some time in his life. Designing directly with machinery that is fully understood – and earning a living with it – can be an education in itself.

The position of the small workshop has been open to attack from every quarter – most obviously as an apparent anomaly in an increasingly industrialized society. From a design standpoint many such workshops produce sentimental, derivative, and sadly out-dated work. The craftsman concerned may be an admirable person, but one who has failed to come to terms with the conditions of today – either intellectually or emotionally – and has often rejected machinery outright, looking backward for his models to (supposedly) kindlier times. It is interesting that such people can see some of our social wrongs very clearly – especially wrongs in working relationships – but that they may be quite unable to see the spiritual complacency in their own work. When machinery is reluctantly accepted, the work itself maintains its high standard of craftsmanship, but has little else in its favour. A few boatbuilding yards head the small list of exceptions, but boatbuilding is anyway a special case. Our crafts exhibitions lend sad and repeated testimony to this failure of nerve in the smaller workshops.

Under these circumstances, architects and other designers are unlikely to take the craftsman-designer very seriously. This is a pity, because much good work could be done by small workshops, and indeed such shops could develop the most rigorous standards of advanced design, together with a personal service and good craftsmanship.

The practical problems must not be underestimated. The most serious is the need for adequate working capital. Although the Rural Industries Bureau (the word 'Rural' is something of a misnomer) in

England and comparable agencies elsewhere, can advance loans on favourable terms for machinery in suitable cases, the beginner is apt to overlook the need for reserves against slow payment, the costs of advertising and transport, the costs of time spent on book-keeping and seeing clients, and the many other factors that may tip the balance between ruin and success. The problem of outlets is an important one. A few workshops have been able to run their own retail shops, which has the advantage of building up a clientele from direct contact, and may be useful for some of the production items that will be necessary as bread-and-butter lines (the jam, however compellingly 'experimental', will be thinly spread, in any workshop that has to be self-supporting). The problem of locale is a difficult one. Workshops must keep in touch with their sources of work, yet city premises are usually expensive. This factor may be balanced by the transport costs of an out-of-town workshop. Beginners usually underestimate floor space, particularly space necessary for assembly, packing, and finished-job storage (particularly in the case of three dimensional work), and it is hard to predict just how much time will be given to finishing and packing in relation to machine and assembly processes.

A further difficulty, concealed to the eye of the hopeful beginner, is that of too much work coming in at the wrong time. Turning down work makes for bad relations, delivering it very late is considerably worse, and rushing the job at cost to technical standards is worst of all. Under these circumstances, it is difficult to find a working balance between a properly equipped and economically viable unit, and its necessary escalation into small-scale factory production, with all the economic and personal pressures that this implies. Some factories have begun in just this way, and have continued to maintain conscientious design standards, but at some cost to the original intentions of the designer. A good designer *can* be a good business man, though few spread their talents in that direction.

Workshop designing can never be a prosperous way of life and will always be extremely hard work. Against this, the experience will be educative, good work can be done, and a small living well-earned. Such shops (at present rare, when operating with a lively design consciousness) could form excellent places for a design apprenticeship.

4 The nature of formal design education

'Well building hath three conditions: commoditie, firmnesse, and delight' – *Vitruvius* (para. *Sir Henry Wotton*).
'Love, work and knowledge are the well-springs of our life. They should also govern it' – *W. Reich*.

A design capability proceeds from a fusion of skills, knowledge, understanding, and imagination; consolidated by experience. These are heavy words, and they refer to the foundations. We accept a certain minimal competence as the basis of professional self-respect, and as some guarantee of a designer's usefulness to other people. Within limits such a competence is definable, and will begin to form outlines within a formally structured teaching/learning situation. It is too much to say outright that design ability can be 'taught'. As with any other creative activity, it is a way of doing things that can only be grown into, perhaps – but not necessarily – in the context of a formal design education.

This view is readily conceded for something as immaterial as 'imagination', but it is commonly held that skills and knowledge must not only be taught, but rigorously examined: if only to protect an unsuspecting society against social or technical malpractice. Defensible as this may be, it is not an assumption that should go unquestioned, nor deflect attention from the weaknesses of received professional standards. The cost of knowledge used without understanding is merely difficult to measure: it is not less real for that. A skill may be irrelevant to the nature of a problem, or – in dealing with people – may be grossly uninstructed in a necessary tact and discernment. Knowledge may be thinly experienced as a rag-bag of conventional responses helped along by access to someone else's published working details. Plainly, skill and knowledge cannot be weighed out by the pound, and separated from qualitative perceptions, for any but the simplest mechanical problems – and even there it is questionable. Even 'judgement', that wise old word, becomes ponderously inhuman unless fertilized by some order of creative spontaneity.

Architecture is only one profession that offers uncomfortable testimony in these respects. It is nice to know, for instance, that a building is unlikely to leak or fall down (in fact someone else has probably done the calculations) and in a simple way this must be counted a social gain. On the other hand there are very many thousands of architects all of whom have passed their examinations, and can we decently say that more than a substantial minority produce buildings of affective quality? Of course their work is sometimes exceedingly difficult and subject to every conceivable restriction, but does their education help them to feel desperately a gap between promise and fulfilment, and thus to find every conceivable way of bridging it?

Let us answer that the best schools know this problem backwards and do their best to resolve it. A 'profession' can still become a self-protection society with a very short term view of the priorities for professional competence. In the long term, we have no yardstick for the spirit of man and the nature of deprivations in his environment – not only the wilderness he may see and accept all around him, but the very nature of his interaction with a wholeness of experience, of which a built environment is but a part.

These are large thoughts, and will surely bring a smile to the face of any over-worked architectural assistant, aware as he is of the drudge component that occupies so much of his conscientious labours. Yet to evaluate formal design education it is necessary to ask some awkward questions. Not only to disturb our unthinking acceptance of social norms, but to bring some very practical matters sharply into focus, and others to dismiss as marginally relevant. For instance, our assistant would confirm that much of his work is elaborately interwoven with building contingencies and the structure of 'consents' which sometimes keeps him awake at night; the personality of the local District Surveyor may occasionally seem the most omnipresent factor in the whole of the job. It is impossible to explain this properly to a student in a school; you simply need the experience of design practice to see how it happens, and just what you do to keep the job moving and your first intentions reasonably intact, or, as is often the case, subtly changing as new possibilities reveal themselves. The way in which accidents of site contingency suddenly appear as benefits, so the designer wonders why he hadn't thought of that in the first place – to explain *that* as part of the 'design process' is really not at all easy.

It is also desirable (to put it mildly) to see the educational problem in terms of the future as much as the present – and, in rather a different way, in terms of the past. It would be a mistake to presuppose a static social situation, pleasantly unified in the untroubled pursuit of affluence, disturbed at most by some new concession to the good life announced in the weekly journals, and the designer providing his expensive austerities where they can be afforded; namely, in the places of high financial decision-making. Our children may take that one to pieces before the last Barcelona chair is glued into position; indeed, there are signs that they are already doing so. Yet a measurable standard of living is not to be despised because not all of us have it or want it, when millions of people desire above all some alleviation of their physical poverty. Nor can we play God and start back from square one with a wholesale redistribution of resources and an imposed system of moral absolutes to keep everything tidy. In short, the designer, like other honest citizens, will need access to faith and vision as much as a keen analytical intelligence – if, that is, he is to engage with life effectively and make something good out of his work. Aside from the specific content of a design training, what can a design school do to help him?

A first requirement for students is knowledge of how they can best help *themselves:* in this respect it is useful to understand the limits and benefits of an academic situation. To begin with, students should realize that both education and design practice are too often handicapped by identity-fixations. The words by which people describe themselves – architect, graphic designer, interior designer, etc. – become curiously more important than the work they actually do. In one respect this is fair, because under modern conditions it may be very difficult to find one word to identify their work, but such words tend to build up irrelevant overtones of meaning which are more useful as a comfort to personal security than as a basis for cooperative enterprise. Such confusions interpenetrate with status values and the other intricate strands of our social life, so it is hardly surprising that education is affected by them. Thus it comes about that design education is often still irrationally divided up into specializations with a doubtful relation to the work students may finally do, and with even less plausible reference to the situation *as it could be* in ten years' time. The fact being, of course, that our design consciousness is still lagging seriously behind its potential, which includes an ability to forestall some of the conditions of an industrialized society, rather than tagging along in the hope of clearing up some of the mess. Human beings have a tendency to shelter behind the familiar rather than confront – or anticipate – the results of their own inventiveness. It is easy to condemn, and easy to simplify this problem, but when Colleges are still talking about two and three dimensions in 1969, and keeping students confined to such categories for a design education, it is certainly easy to feel very frustrated indeed.

This problem is only in part a structural one; it cannot be solved by denying that specific disciplines relevantly exist, nor that they are worth studying in depth. In theory a three-year course could investigate the Universe from the confines of a single problem. It is also a fallacy of bad technical training – as implied earlier – to suppose that skills and knowledge can be picked up *in vacuo* or in neat packages as in a supermarket; such knowledge is sometimes used with a contempt and a restlessness that betrays the additive nature of its learning. Skills should emerge organically in a student's experience. Again, lateral thinking must be complemented by vertical thinking; which usually happens best in one place. Much design work can be approached with freshness and insight with very little in the way of equipment and materials: the assumption to the contrary (as an accepted starting point for design) merely reflects a social climate in which the least meaning emerges for a gross expenditure of effort and apparatus. Thus a restless and dissatisfied student who seeks his freedom in the extent, rather than the depth, of his explorations, may be dancing to the tune of the masters he most decidedly wishes to replace.

The case against education in 'fixed' categories is quite strong enough to accept such considerations as open to test and experiment. First, the categories are usually arbitrary and may not even refer to the professional realities of today, apart from the predictable needs of tomorrow. Second, students are various and variable human beings; their needs change and were not the same when they entered the College. Third, some design problems can benefit from technical exploration in more than one area – and from the conceptual gains which may ensue. It is not impossible to provide accessibility between adjacent disciplines and technical facilities, though sometimes it is not easy to do this properly. The need will always seem more pressing and insoluble in schools that resist informal arrangements, or in schools that have no clear and intelligible academic programme backed up by good personal relationships. It is a sad but interesting observation that the most rigidly blinkered courses of 'training' are often the least *technically* competent, rather in way the that some committed amateurs do better work than tradesmen who have lost a responsible relation to their trades. (*Amateur* – one who loves . . .) However, there is a radical need to appraise the common factors in every design specialism and provide for their study by all design students working together – for example, communications theory and practice, ergonomic and psychological studies, problem-solving work. This provides the basis for an understood (and subsequent) bias toward any given specialism. The supplementary benefits should be self-evident. This is not the place to go further and prescribe the outlines of a universal design course; each course must draw its life from local and actual circumstances. It is really not difficult to 'design' a course that can stand up to argument, exploit local constraints, accommodate necessary variables, and gain assent both from students and the design profession – though, considering this as a design problem, there is much variation of quality in the way the problem is both stated and solved. But the best theoretical framework for studies will be alive or dead rigid, according to the spirit in which it is interpreted, both by students and staff, which in turn is a function of confidence within the academic community.

There are more subtle considerations that students should keep in mind. The period of 'further education' is a critical time of life, rich in subjective discovery and active in the growth and change of personal values and attitudes. Students are unusually thick-skinned if they accept such experience placidly and without anxiety. A design school may be in some sense 'preparatory' for life to follow, but a student's time there is precious and irrecoverable life in the present. It is also, however, a safe berth in harbour. In a somewhat different way, the staff of a school seek their own expression in the achievement of the school; it is their own chosen and continuing way of life. On the whole, the work is agreeable, relatively well paid, and socially well regarded. It is work that allows some teachers to defer indefinitely a close look at their own inadequacies. For many reasons – not all so uncharitable – an educational *connivance* becomes possible between students and staff which may give a wholly false emphasis to the importance of structured education. There is a good sense in which any designer worth the name will be a student for the whole of his working life. This is not only a function of creative resource, but also of the conditions of rapid technological change which a designer must meet in his work. In such a perspective, the few years in a design school are not unimportant, but should be carefully guarded against inflated claims for 'completing' the education of a designer. At best, three –

or even five – years in design school can attend (thoroughly) to a few simple priorities in a designer's personal education. Even at a mundane level, much of what is called 'operational know-how' is necessarily picked up in the rub of professional practice.

The value of a design school emerges from the fact that numbers of staff and students are gathered together in one place with a common purpose and common facilities. Students learn a great deal from each other; not only from books or from their tutors' guidance. These facts point to the benefits of academic life: a gradually widening area of agreement (the norms against which individuality becomes meaningful), the experience of sharing, co-operating, and resolving conflicts: in a word, the chance of *participation* in all the stress and stimulus of a particular community with shared aims. This side of 'further education' is no less formative than the acquisition of skills and knowledge, much of which (not all) is equally available to the student who teaches himself from books, correspondence courses, or from the hard lessons of practical experience as an apprentice or design office assistant. The reality factors in academic life may appear to derive less from studio project work, which is always imperfectly 'real', than from the discussion that surrounds such work, the exploration of angles of attack, and the slow take-up represented by the experience of community.

Design education *must*, by its nature, dig below the surface, and must at the outset be more concerned to clarify intentions than to get results. If it is sensible to see learning and understanding as rooted in the continuum of life, it may be that a really useful introductory course will only show its value in the full context of subsequent experience; i.e. several years afterwards. Conversely, an education that concentrates on short-term results may give a misleading sense of achievement and fail to provide an adequate foundation for subsequent growth. This is a thorny problem, because under the pressurized and success-conscious conditions in which we live, students are naturally anxious to prove themselves as rapidly as possible (to themselves and their contemporaries and teachers). Something as intangible as the growth of understanding may seem a poor substitute for the almost measurable achievement marked by a high output of design projects, however specious or thinly considered such projects may be.

Diploma work is a serious business for students, and can cause a great deal of anxiety, particularly to those who have a (perhaps well-founded) dread of examinations in any form. If such students will realize that their school work is but an intensive phase of a very lengthy design education, perhaps they will feel less undermined by the pressures of Diploma work. The purpose of coming to a design school is *not*, primarily, to gain a first-class Diploma, but to make a constructive use of several years' education. Some students, on the other hand, will find the edgy business of first and second-class Diplomas a positive stimulus or a useful objective to work toward. This is a matter of individual psychology: in a good school the course work should be rewarding enough to keep such matters in a tolerably *angst*-free perspective. An intelligent design course will also recognize that much design work is shared cooperative effort: students will be encouraged to help each other. In such matters it is sometimes necessary to educate the educators. A small-minded and up-tight view of human experience will take small differences of accomplishment very seriously. A more generous view will allow that we are all holding candles in the dark.

The obvious defect of academic education is its remoteness from the conditions of 'real life' design practice. Simulated project work is always elusively different, however well set-up. (In passing, it is a *sine qua non* for academic projects that the brief should be scrambled, as it always is in design work; that its vagueness should then be intensively questioned by the student; and that it should finally emerge in reconstructed form as very clear and precise terms of reference.) There are good reasons why design courses should assimilate a fair proportion of live jobs into the academic programme, but there are also administrative and academic difficulties that would take two pages of this book to analyse. Some difficulties simply reflect bad habit and a lack of initiative in a school; others point to the unwarranted time that a single live job can occupy in a course, the need to balance and integrate studies, and so forth.

If this situation is approached in a negative spirit, a school will lose much of the fun, enterprise, and commitment that might derive from lively relations with the local community. In every city there are projects – schools, hospitals, play groups, welfare organizations – that are often desperate for help and equipment. The conditions for 'success' are not nearly so critical as in ordinary commercial work – except in the case of special purpose equipment (say) for hospitals that demands a research programme. Even here the design schools should be making a contribution. Every school should have its own design office in which teaching staff are kept creatively active for a part of their (paid) teaching time, and in which students can work – in effect – as apprentices, during a part of their design course. This is a very fast and practical way of learning. Projects in which students work both as designers and executants should be chosen with some care, because students usually underestimate the time spent in supervising and completing a job – something, on the other hand, that they might well discover for themselves. Schools might also run a local shop; indeed, there is no limit to the ideas that might emerge from a socially committed view of education, as distinct from the view which directs effort toward window-dressing and constant onerous assessments of a student's supposed 'progress'. The possibilities can be discussed and related to academic objectives, *if* there is free discussion at all times between students and staff, and if someone has taken the trouble to define and argue for the content of such objectives.

This implies goodwill and intelligence from students, not just staff. Any student who is still smarting from the effects of a bad secondary education, or who has difficult personal problems, will always tend to project deep-laid resentment on to his future relationships in the education field (and elsewhere). This is a perfectly ordinary problem in any kind of further education and must be met with every possible form of help, including if necessary the provision of noncourses for people who really feel they have been pushed around too often.

No student should claim total exemption from the effects of his background and some twelve years of structured school education, with its own power to liberate or depress. This is noticeably a source of anxiety in art schools (assuming here that a design school is a component part of an art school). A proportion of students there may be in full flight from a secondary education that has been too rigid or competitive for them, or which has failed to provide a balanced growth of their faculties, or which, in failing to gain their confidence, has inhibited a valid appraisal of the alternatives in further education. 'He's good with his hands but . . .' is only the extreme of this deprived situation. Thus some students may be questionably placed in their choices, by any specific criteria of talent, although the 'feel' of the art-school environment may be wholly right for them. In practice there does tend to be a therapeutic component as a recognizable function of art education, whether or not it is accepted with good grace. A ghetto system separating out 'fits' from 'misfits' is hardly likely to benefit either group, nor is the separation so apparent. An art school has to be a place where people can find themselves at their own pace, and schools which provide little room for manoeuvre will carry the burden not only for their own shortcomings, but also for those of the whole education system.

In design schools this problem may be seen a little differently, because the work is less personalized and subjective, more work is necessarily done in groups, and there are better and more frequent opportunities for discussion as part of the normal way of working. In both cases the educational requirement is not so much for a blanket 'permissiveness' to quieten down anxieties about life in the twentieth century, as for a community of honest discourse in which such words simply do not arise. A 'free' design school is not one in which nobody cares enough to find out that good work is difficult, or that the life and service of ideas can involve a painful apprenticeship. A teacher who suppresses his own convictions, or creative experience, in pursuit of a notional value of the general good, is too often a man of tepid enthusiasms but good intent; his opposite number, the bully or bigot, may prove a turncoat and prefer to join him in a climate of flabby nonconformity. An active critical tradition is more taxing, and it needs the 'critical models' without which discussion becomes self-nourishing and dies.

In a practical way, a free school must tend the variables in its situation, offering an ultimate self-determination to the students and every possibility for growth in its own structure and teaching approach. Yet central to its traditions must be the experience and analysis of necessity (something of which is discussed further on). There is something else that may be overlooked. Teachers must be helped by students to learn from the work and the relationships they encounter, thus rescuing their position from a progressive loss of confidence which may well find expression in defensive or authoritarian attitudes.

Teachers who are practising designers can also lose ideas which may have cost them years of work, assimilated instantly and almost unnoticeably in the ordinary way of teaching exchanges; only to meet years afterwards the reproach that they have exploited their students' ideas to their own advantage. The origin of the situation is long forgotten, but if a designer is teaching for any long length of time, he may feel increasingly deprived of his own creative identity.

This is the case for short contracts and an extension of designing opportunities for teachers in design schools. It also argues for the apprenticeship system as a component in education, because however careful a teacher may be in avoiding the intrusion of his own design approach, and however questionable the concept of 'loss' in this context, design teachers will know that this problem is a real one (and students may feel that it can operate clearly enough in reverse). Such difficulties become obsessive only in over-structured teaching relationships, where the us-and-them structure requires guarded attitudes akin to property-relations, and in which students and staff fail to discuss informally their respective roles.

In this subtle matter a young student, with all his relative inexperience, can give as well as take. Freedom – acted out, analysed, reflected upon – is the tyranny and the necessity that can help to foster such perceptions. Student direct action is a special case – see remarks to follow – but those who joined the students at Hornsey or Guildford (England), in their protracted sit-in, will confirm that here was a laboratory situation for experiment in relationships, one in which the old (had they been there) would have found much to learn from the young. As it was, all were students together; there was no other possibility. A man who had experience to offer was respected for his contribution, and that was that. To argue from the floor for what you believe in, is to see how much nonsense passes for 'responsible judgement' when elevated on to a platform.

Specific decision-making is sometimes rejected in favour of a weakly metaphysical notation of 'process', simply because students cannot stomach the second-rate (as felt, rather than evaluated) and they may have seen little good teaching or good design either. There are times when to say 'no' is a constructive act; to say yes, *as a designer* looking to the future, is to join social commitment to a mastery of particulars. The smallest tangible accomplishment is not to be despised. Teaching that helps to sharpen anticipatory skills (on present realities) has little need of platforms and robes of office. A good design teacher may be defined as one who can communicate both his areas of confidence and the limits of his own awareness, so that he is able to advance the fruits of experience in a spirit of positive uncertainty. In this discretion, and in this spirit, 'models' and 'constructs' may become acceptable and unrestricting. As used here, these words refer to verbal or visual analogues to design situations (models) and hardware that can be touched, looked at, and performance-tested (constructs). They can be seen simultaneously as projections from an experienced response, and as the fruit of personal insight which must remain provisional. If these are obscure words, then Frederic Samson (addressing students at the Royal College of Art) puts the matter pithily and with characteristic insight: 'the main difference between us is that your ignorance is superficial but mine is profound'.

These are personal considerations; the meeting of people face to face in the realization of some shared purpose, the 'half chosen' nature of which is warning enough. It remains to add that the urgent practical problem is making *more* meetings possible; cutting out the red tape to encourage architects and designers of all kinds to come into the schools and talk informally about their work. This is a high priority on the budget of any design school. The wider and more theoretical problem hinges on a detected conflict between immediate social requirements – the training of 'marketable skills' – and the role of education in refreshing social values, and thus helping to define social ends. There are difficulties here, but not an essential dichotomy of aims that cannot be reconciled. Even on the most conservative view, the principle of culture-continuity, and of culture transmission for which formal education is the most ponderous vehicle, ceases to have classical validity unless education can keep ahead of the situation it exists to serve. When it does not, the result is apathy, a feeling that life has in some sense 'gone elsewhere'. In extreme cases there may be a complete breakdown of communications to the point where students' direct action takes over. The issues at stake are usually real enough, but they symbolize much wider and often formless misgivings.

In the search for 'a correct pedagogic approach' – perhaps rather a quaint concept – the Hochschule für Gestaltung at Ulm has researched with energy and seriousness. This school has established a polarity of thinking and attitude that merits close study. Tomas Maldonado is quoted as saying: 'Education for design has become a very complex task. We must train people capable of revolting against stereotyped ideas, but we must also equip them with the means to do this; otherwise the result is merely declamatory. Moreover, in most cases the act of creation is not something beginning and ending in an individual. It is a social fact.'
Out of context, there seem to be buried assumptions in this statement

which give an uncomfortable fixity to its viewpoint – though the notion of declamatory thinking (and work) is a valuable one. Must the problem be seen in so paternalist a way and what are the conditions for the credibility of social fact? What needs are satisfied and hungers alleviated?

The challenge of a fast-changing global view of life stretches the imagination like elastic; it is ever more necessary that experiences nearer home are tethered to credibility. I would dwell on this need of credibility, but I think students, including those long past their formal apprenticeship, will know what I mean. There is certainly no doubt that many institutions have become seriously over-structured for the job they have to do. In education and elsewhere, the grip of outworn forms may discourage new energies beyond the toleration-threshold within which differences are normally met and resolved. The parties involved find themselves talking in different languages – in the same tongue, but with radically conflicting assumptions. Such new energies may be quite unformed, supported less by argument than by exploratory behaviour as such, and may be reaching toward new insights along unfamiliar paths of social exploration. This is the genesis of the student demonstration or sit-in. It is short-sighted to regard these happenings as essentially destructive, when they may generate and focus an extraordinary energy of self-education, quite remote from the ostensible aims of 'protest'. Indeed it is possible for those who detest revolutions to see such actions as genuinely an attempt at social adaptation; that is, to conditions which more ordinary channels of expression are failing either to contain or to anticipate. (Traditional forces of reaction will doubtless learn to keep up with the times.) Certainly a husk of discarded usages is revealed on these occasions, and one that has failed to grow with the perceptions of an impatient minority. At worst, the reaction provoked is too violently defensive to promise more than deep disillusion to the revolutionaries. At best, an institution is jerked out of complacency into a new consciousness of its privileges, and, for the participants, there may be a deeply moving experience of those twin principles for human conduct – solidarity and reciprocity (all for one and one for all, and an active spirit of empathy in human relations). The result is not (as commonly supposed) a developing group hysteria, but may be a notable growth of self-knowledge and individuation.

For such experiences to 'take' and bear fruit, they should be seen and valued for what they are – stripped of rhetoric and accessory content, and related to a continuous search for productive working attitudes beyond the confines of formal education. For students committed to an immediacy of experience in their day-to-day lives it is not easy to think in such terms; reason enough for practising artists and designers vigorously to support every significant breakthrough against academic inertia. Nothing can seem more heartless and desolating to young people than the slow erosion of vision by an aftermath of patchwork compromise. Yet to grow along with this suffering – it is that – and to turn it without bitterness to good account, is merely to share the predicament of every lover of life.

It is reasonable to mistrust any theory of conduct which draws life from the presence of an external enemy (or the necessary creation of such). Yet there is a dangerous doublethink that permits violence and apathy to be institutionalized in society, whilst deploring the faculty that young people have for detecting humbug in their elders. The intelligent way to meet any promise of new life must be with gratitude untinged by cynicism. Given favourable conditions for experiment (including acceptance of the premise that this century urgently needs experiment at every level of social organization) then the worth and staying power of new ideas will swiftly demonstrate themselves. Above all, the human need for pluralistic solutions must be respected against the threat of some new and paralysing orthodoxy. Taken neat as a prescriptive cure-all, with no awareness of local conditions, even a 'network' can become a cage of emptiness.

In design education, two modest lessons may emerge helpfully: the need to connect apparently unrelated phenomena, if fresh insights are to accrue; and the reminder that design presupposes the setting up of criteria for value judgements. Every design student knows these things (or should do) in the ordinary context of problem analysis. It is another thing again, however, to see the whole of our physical environment as an intricate pattern of choices, few of which are inevitable right up to the point of their hardening into social fact. For some such choices a designer can exercise more than a formal responsibility.

Formal *education* must certainly give access to the options that face us all and that make, at best, a shadowy appearance in the careers pamphlets. If the gap between technical capability and social imagination is to be bridged, it seems likely that education must lose much of its formality to seek far more warmth and flexibility of outlook. Every single concept of structured education may need to be reconsidered in the coming years, with perhaps the realization that the informal truths are closest to sources of wisdom and creativity. There will be failures again and again, but there is a difference between failure by default, and failure in attempts to do something really worth-while. A free school may be as elusive as a free society; it is the effort to build both, and bring them together, that could form a truly comprehensive education.

As for design, all we can do is make good work possible, and be alert to its coming; never fooling ourselves that all good things come easily. To work well is to work with love. A hail of words, like rain in April, can do no more than keep the air sharp and sweet and the ground springy underfoot; and that is the best a formal design education can hope to do – relevantly.

The Artist

The artist: disciple, abundant, multiple, restless.
The true artist: capable, practicing, skillful;
maintains dialogue with his heart, meets things with his mind.

The true artist: draws out all from his heart,
works with delight, makes things with calm, with sagacity,
works like a true Toltec, composes his objects, works dexterously,
arranges materials, adorns them, makes them adjust, invents.

The carrion artist: works at random, sneers at the people,
makes things opaque, brushes across the surface of the face of things,
works without care, defrauds people, is a thief.

Toltec verse translated from the Spanish by Denise Levertov

5 What is good design?

The 'goodness' or 'rightness' of a design cannot easily be estimated outside a knowledge of its purpose, and sometimes also of its circumstantial background. This is no reason for timidity of judgement; a man must reserve his right to say 'I like that; to me it is beautiful and satisfying, and more so than that one over there that works so much better' – or, 'this is a good workmanlike solution, thank God it has no pretensions to Art'. Theoretically, a well-integrated design should come so naturally to eye and hand that neither of these comments will be called upon, but human nature isn't so simply natural and nor is human society. An optimum solution is possible where the conditions for verification can refer to absolutes; a daunting and illusory requirement in most design situations. On the other hand, a design can say to us 'here is a problem that is so well understood that it can be felt to be moving toward an optimum solution; the design is inclined in that direction'. This is designers' talk; the user of a product will not be too interested in the skill with which a designer has met his constraints. If a design is so well wrought that overtones of meaning are present, so that the work can be experienced (optionally) at many different levels simultaneously, then it is a condition of *organic* design that the further harmonics must not clutter or deform a simple level of acceptability.

For the designer, good design is the generous and pertinent response to the full context of a design opportunity, whether large or small, and the quality of the outcome resides in a close and truthful correspondence between form and meaning. The meaning of a good garden spade is seen in its behaviour, that it performs well; in its look and feel, its strength and required durability; in a directness of address through the simple expression of its function. More complex objects, places, equipment, situations, may well exhibit less obvious dimensions of meaning – of which one may be the property of reference discussed further on in this book. A design *decision* may prefer some determinant principle of action to a material outcome. As a social activity, the integrity of design work proceeds from the understanding that every decision by one human being on behalf of another has an implicit cultural history. Design is a field of concern, response, and enquiry, as often as decision and consequence. In this sense (also), good design can both do its job well and speak to us.

Every design product has two missing factors which give substance to abstraction: realization and use. These are the ghostly but intractable realities never to be forgotten when sitting at a drawing board. In a similar way, any discussion of design philosophy must never stray too far from nuts and bolts and catalogues and every kind of material exigency: a designer breathes life into these things by the quality of his decision-making. Thus his concern is truly 'the place of value in a world of facts' (*cf.* the book by Köhler of that name) and the outcome can (or should) be a form of discourse; but not a verbal one. His work can be said to deploy the resources of a language and be accessible to understanding through the non-verbal equivalents of intention, tone, sense, and structure, (*cf.* I. A. Richards), but there are other and more directly functional levels of experience which – as has been said – must come to the hand with all the attributes of immediacy. Most of the time a designer finds it hard enough to do small things well. Any number of broader considerations must not distract him from that task, but rather enliven and give sanction to its meaning.

A product must not only be capable of realization through manufacture, but in its very nature must respect all the human and economic constraints that surround production and effective distribution. This may seem obvious in the case of product and communications design. Similarly, it is difficult properly to evaluate a building without some idea of the cost factors and the client's briefing. Difficult, but not impossible; because a clear design will generally manage to state its own terms of reference, unless disaster has intervened at some immediate stage to distort the central intention of the work. There are many cases in which a good design will be discarded for reasons which seem arbitrary, perhaps to be replaced by some meretricious product with a better sales-potential. Again, a perfectly adequate design solution, the result of much care and imagination in its development, may never reach the public at all. In this respect the artisan-designer may enjoy a freedom denied to the designer for mass-production (though his economic problems will limit the scale of the work), and much experimentation in form-giving will necessarily occur in situations exempt from marketing difficulties. These will include one-off jobs, limited production runs, and public work (e.g. schools, hospitals, airports etc.). Much early discussion in the modern movement assumed – broadly for social reasons – that product design (i.e. mass production for consumer market) was the centre of inertia that had to be revitalized: Herbert Read's book *Art and Industry* reflected this assumption in the 1930s.

Gropius's *New architecture and the Bauhaus* contained a classical statement which seemed to imply that product design would move inexorably toward the 'type-form' for the problem examined.

As things have turned out, the most interesting work has happened, of course, where it was economically possible. The domestic consumer market has gained an important component in DIY (do-it-yourself) which in itself demands a reappraisal of the designer's role in the areas affected. The mass-production of building or service components (e.g. pressed-out or moulded bathroom service units) has hardly approached the potential seen for it forty years ago. The notion of *place* as the focus for communal achievement has scarcely fought off the demands of *occasion* and mobility, despite moving and articulate pleas from Aldo van Eyck and others – and despite the continuing reality of place as a factor of ordinary experience, eroded as it is by communications, and the rarity of imaginative work in this field. It is a mistake to see the designer's work as conjuring up new worlds at the scratch of a drawing pen: there are many fields in which the designer could profitably work with (for instance) do-it-yourself and cooperative housing agencies; and there are fields in which a designer can and should respect the organic continuity that surrounds people's lives. Two examples: the interior designer is doubtful of his 'respectability' because every one knows that architects should design their buildings from the inside-outwards. In fact, there are plenty of buildings that are simply weather-proofed and service-provided shells, waiting for specific uses to be provided for. However, leaving that aside, it is not a necessary argument to suppose that – given adequate social resources – the whole of our physical environment should be uprooted and totally replaced at regular intervals. This is a dangerous fantasy. In fact there is plenty of scope for the adaptation of existing buildings to new uses (a so-called slum area is as much a pattern of relationships as of decayed buildings) and this is interpretive work for which the 'interior designer' could be well-fitted. Again, there are plenty of structurally sound buildings that could be given extended life with the aid of a loan and a do-it-yourself handbook. As it is, the lunacy of high-rise development has only recently been seriously questioned; in practice, land values give rise to extraordinary palaces for paper-work springing out of areas of private squalor, and the simple things – like the provision of neighbourhood amenities – are neglected in favour of drawing-board schemes which may seriously debilitate the life that 'squalor' sometimes reflects. A run-down neighbourhood may need a lot of things but the problem must be seen in more than a tidy-minded way; every problem, however complicated by planning and growth statistics, is met with concealed assumptions (and often concealed economics). Here is ground both for humility and for diagnostic sensitivity in the way a designer approaches his work.

The difficulties for product designers are not just a matter of plain villainy on the part of manufacturers; they are, in part, a consequence of capitalism. Whilst strange things do go on in boardrooms, it must be realized that a well-designed product must be sold competitively. Experimental work may be chancy as a sales proposition. As things are, a first duty of a company director is to make his company profitable (which he may conceive as a first duty to his shareholders), and a second duty is to keep his work-people in continuous employment. Experiment becomes a closely calculated risk, very much at the mercy of the buyers in the retail trades, and subsequently dependent on a successful advertising policy, public response, and many other factors. In the furniture trade there are a few companies who have tried to maintain reasonable design standards, against the hope of improving them as the market 'softens' sufficiently to warrant further advancement. Such companies have relied on contract work – furniture for public buildings specified by architects or local authorities – to help carry them forward. It will be seen, at least, that under ordinary production conditions, product design cannot easily be evaluated against absolute standards, yet products meet constant criticism on such terms.

Unfortunately, it is also true that there are innumerable products that are just very poor realizations of a straightforward and entirely non-experimental design concept. They could have been marketed just as easily had they been designed with more distinction. The design capability simply was not there. Designers should be aware of property-relations as a conditioning factor in the way they design (and think about design), but no designer should fool himself that given 'a better society' it would then be magically easy to design *well*. A designer who stops designing in the hope of better things may lose his ability to design anything at all; to this extent people become what they do. Here, an idealistic student might consider the partial truth in the saying 'a few are artists, the rest earn a living' – which in caricature might be said of every profession and not less of the sciences than the arts. Those who elect to put their work before everything else, which is merely one of the conditions for complete mastery in any field, must fairly expect life to present some difficulties.

The hard facts of a market economy are easy to overlook in the relatively permissive ambience of the average art school. Although academic life is subject to its own peculiar stresses, economic sanctions are not pre-eminent among them. Fortunately, students need not harden themselves against a perpetual winter of creative frustrations: the situation is not as depressing as some of these remarks might suggest. It is true that a designer's freedom will reflect in large measure the values of the society in which he works. Designers are not privileged to opt out of the conditions of their culture, but *are* privileged to do something about it. The designer's training equips him to act for the community, as (in limited respects) the trained eyes and hands and consciousness of that community – not in some superior human capacity, but in virtue of the perceptions which he inherits from the past, embodies in the present, and carries forward into the future. He is of and for the people; and for them, and for himself, he must work at the limit of what he sees to be good. The sentimentality of talking down, or working down, is a waste of the social energies invested in his training: thus can 'social realism' enshrine the second-rate.

If society is geared to satisfactions on the cheap, the designer has a special responsibility to straighten himself out in that respect; to decide where he stands. When real needs are neglected, and artificial ones everywhere stimulated into an avid hunger for novelty, sensation, and status-appeal, largely (but not wholly) for reasons of private or public profit; then here is his own nature, his own society. He is involved, and he must decide how best to act. It should not surprise him to find a thin and pretentious reality informing the design language of the world which he inherits. A Marxist (or Anarchist) analysis may be one tool to help him sort this out, but he will hardly need to put on Marxist spectacles to see that a veneer of good taste has 'reference' to certain obvious social conditions and is not the whole of good design. The design student may sometimes find that the industrial scrap-heaps, the surplus stores, and the products of straightforward engineering, will yield images of greater vitality than will be found in more fashionable quarters (though even here, fashion spies out the land). Such a situation is a challenge, and as such must be studied and understood.

Yet it is still no answer to live in the future; every skill must be nurtured by a commitment in depth to the present. The meaning of creativity may be seen as an equation which resolves this apparent paradox. Work that lives is rooted in the conditions of its time, but such conditions include awareness, dreams, and aspirations, as much as the resources of a specific technology: such work respects the past and actually creates the future. These problems, and their wider implications for human happiness, will necessarily concern students of design, because no one can make truly creative decisions without understanding; and without a real participation in the constructive spirit of his time. *This spirit must be sought out*, not necessarily by intellectual means, to be honoured wherever it is found.

Students who are depressed by the visual poverty of our surroundings should study the spirit of the 'modern movement' in its development from the turn of the century to the late 1930s. Here they will find themselves in good – and various – company. As Walter Gropius put it, the modern movement was not a matter of 'dogma, fashion, or taste', but a profound wide-ranging attempt to encompass the nature of twentieth-century experience and meet its physical demands with a constructive response.

What may excite us most obviously about this phenomenon is its tangible achievement; a whole word of very explicit imagery willed and impelled into being, almost – as it seems in retrospect – from zero; a new beinning. The fact of historical conditions very different from our own does not diminish the marvel of this achievement and its continuing relevance to our own position. This relevance has little to do with the forms that emerged – and nothing to do with imitating them. It is important to realize that much 'modern' design, and particularly what is known as 'contemporary', is an enfeebled and misunderstood derivation from this earlier work, almost wholly removed from the force of its guiding spirit.

Anyone who sees the modern movement in stylistic terms will fail to understand its essentially revolutionary nature. The present situation expresses a frustration of consciousness (and of achievement) and leaves us with a three-way split from the hope of an organic tradition: in the centre, the world of establishment design which some students find so bafflingly effete, and the breakaway into the neutrality of a verifiable non-aesthetic methodology on the one hand, and a fantasy world of a continually receding plug-in future on the other. This is within the professional world; outside it, the interesting mix of DIY, pop, and kitsch which for all its surface vitality is deeply compromised by infantilism and commercial exploitation. The modern movement is best understood, not by reading the books but by trying to join hands with its spirit in your own work. The implications will soon become apparent – relationship not self-sufficiency, every part working for its living, no unseen props, verification from a shared attitude of exploration and from number, search for solution within the problem, the most meaning from the least effort in material resource, clear expression of functional relationship, production for need, not profit . . .

Something of this is well expressed by Paul Schuitema, a still-active graphic designer who worked in Holland and Germany in the 1920s and 1930s.

'. . . We didn't see our work as art; we didn't see our work as making beautiful things. We discovered that the romantic insights were lies; that the whole world was suffering from phraseology; that it was necessary to start at the beginning. Our research was directed to finding new ways, to establishing new insights – to find out the real characteristics of tools and creative media. Their strengths in communication – their real value. No pretence, no outward show. Therefore, when we had to construct a chair or a table, we wanted to start with the constructive possibilities of wood, iron, leather, and so on; to deal with the real functions of a chair, a living room, a house, a city: social organisation. The human functions. Therefore, we worked hand-in-hand with carpenters, architects, printers, and manufacturers. . . . To reduce chaos to order, to put order into things. To make things more clear, to understand the reasons. It was the result of social movement. It was not a fashion or a special view of art. We tried to establish our connection with the social situation in our work . . . The answer to our problems must be the questions: why? what for? how? and with what? . . .'

The attractive qualities of this statement should not blind us to the fact that the modern movement was a minority struggle, carried on against a good deal of practical opposition, and at best, a widely felt social indifference. At least the conflict was capable of clear definition, even if, for the purpose of these remarks, the words 'modern movement' are little more than convenient shorthand for a complex phenomenon with its own contradictions.

It is necessary to stress some of the background considerations which prompted modern design into being, because it is too easy to study the designs that emerged as specially privileged historical monuments, whereas the spirit that conceived them is still alive and accessible to us. In forming our own criteria for 'good' design, we cannot, of course, escape the half-conscious assumptions which make us always the children of our own time, but we do well to remember that our own concerns are in some respects closer to the pre-war period than to the world of the 1940s and 1950s. A whole complex of emergent ideas, values, and experimental work was traumatically cut short by the experience of Fascism, the horrors of Auschwitz and Hiroshima, and by the slow aftermath of cultural assimilation. Not only were energies dispersed in a practical way, but their foundations were uprooted. The implicit philosophy which underpinned modern design was never very far from what is wearily referred to as 'a rational view of man's conduct': the hope and even the confidence that if technology could only be integrated into meaningful value-structures, a new and fruitful way of life lay open to man's willing acceptance. The last war brutally damaged that hope. A rational view must examine motivational forces with a more intimate sense of their origins, and the cost of their frustration. In the design field it is not a matter of exchanging effective imagery for austerities that have had their day. A language of gesture and exclamation tends always toward infantilism; a measure of its warmth but also of its inadequacy. If a new synthesis of thought and feeling is to be attempted, we must think and feel our way toward the place of design in a necessary context of social renewal. Nor must we forget that a warm heart and a rather special view of history do not make a designer. Designing is very specific; a cultivated understanding is no guarantee of a specific creativity. This is the individual problem and a central concern of this book. For the social task we have fresh evidence all the time of man's fallibility, of his deepening technological commitment; of the nature of affluence divorced from social or spiritual awareness. Yet there is a pedantry of the spirit in dwelling too much on these things. The force of new life can break through where and when we least expect it; as in Paris in May 1968, when the impossible seemed suddenly within reach.

It should at least be clear that to speak of 'good' design is to speak of, and from, the conditions of our own time, and our response to these conditions. The intelligibility – and perhaps the existence – of a design 'language' is a problem of the cultural fragmentation that affects participation in every other aspect of our culture. Because the realization of a designer's work is always socially contingent, his freedoms are always a recognition of necessity in a most explicit way. An elegant design solution is one that meets all the apparent conditions with a pleasing economy of means. A fruitful solution co-opts the conditions into a new integration of meaning, whereby what was 'apparent' is seen to have been insufficient. Such answers have questions in them.

For the rest, we must hope, and work. By developing to the full his professional skill, a designer's advice can command social respect, and he can further the art of working *with* – as well as for – his client's own interests toward a wider view of the social interests he is serving. In strictly professional terms, the contribution rests in this accord with society; a widening recognition of services honestly and skilfully given. In personal terms, a designer can decline a commission if it seems to him a waste of time. Life is too precious to squander in that way. At work, every student who has the tenacity of spirit to hold firm against shoddiness and triviality can find his way to new definitions of good design in the social role of a designer. The task is not easy; the temptations to a shallow opportunism are everywhere apparent, and the real achievement may seem small. There is nothing new in the nature of such choices.

6 Summary

Students who want to be useful in the world, who see chaos and want order, who lack self-confidence, and who are swamped by the apparently required armoury of skills and facts, are sometimes rashly seduced by unitary views of disparate phenomena – it seems hopeful that way, and it seems manageable. Those who court lateral thinking to infinity can easily arrive at the semantic doublethink in which concrete equals abstract; an achievement which some design theorists have managed effortlessly. Many a pretty paradox can flourish, and indeed be artfully cultivated, in a society that lacks a spiritual dynamic, and would like the best of east and west without the penalties of either.

For students, the air hisses out when College finally closes its doors and opportunity becomes both concrete and immediate – and small. Dolci used to say that the ones who really helped in Sicily were those who could do small things well, not the cosmic-eyed visionaries with eyes out of focus. Hundreds – thousands? – of architectural students have circled slowly back to the (patio'd) semi-detached, or the spec. building work, that they fought off so bitterly during those golden days of preparation for an inflationary tomorrow.

Design has a very broad spectrum of opportunities that are (can be held to be) socially worthwhile. Some are mechanical, some ephemeral, some interpretive, some call upon unique solutions, some move toward those generalized type-forms that Gropius talked about. Designers themselves are equally various in personal make-up. The first task for a student is to know – and accept – himself. The second is to educate *himself;* accepting formal education as one set of constraints and opportunities exercised within the temporary benefits of community. Remembering, in this matter of education, a saying by Robert Graves – that it is easier and more common to hate hypocrisy than to love the truth. The third task is to sense out, quite intimately, a growing sense of accomplishment that is accessible simultaneously to eye, hand, reason, and imagination; testing each ground to spring from.

This is not an original specification for survival. The point to reiterate is that so few survive at the level of creative expectancy that formal education generates – and to which subsequent experience adds, too frequently, a tepid aftermath. As things are, the one thing about the future of which a student can be sure, is that its demands on him are strictly unpredictable: the rough shape of possibilities may be discernible, but the exact and tangible nature of a creative challenge is a happening and not a forecast.

How, then, to keep open to the future with enough personal acumen and buoyancy to cope with its opportunities at full stretch? If students feel blocked by society as it is, then they must help find constructive ways forward to a better one. In a personal way, the question *must* be answered by individual students in their own terms, but as far as design goes, it is possible to see two slippery snakes in the snakes and ladders game. The first snake is to suppose that the future is best guaranteed by trying to live in it; and the second is an assumption that must never go unexamined – that the required tools of method and technique are more essential than spirit and attitude. This snake offers a sterility that reduces the most 'correct' procedures to a pretentious emptiness, whether in education or in professional practice.

The point is reinforced by another consideration. There can be a certain hollowness of accomplishment known to a student in his own heart, but which he is obliged to disown, and to mask with considerations of tomorrow, merely to keep up with the pressures surrounding him. Apart from the success-criteria against which his work may be judged, there is a more subtle and pervasive competitiveness from which it is difficult to be exempt, even by the most sophisticated exercises in detachment. Hence the importance of recognizing that education is a fluid and organic growth of understanding, or it is nothing. Similarly, when real participation is side-stepped, and education is accepted lovelessly as a hand-out, then reality can seem progressively more fraudulent.

Fortunately, the veriest beginner can draw confidence from his eighteen years, once it is realized that the foundations of judgement in design, and indeed the very structure of decision, are rooted in ordinary life and in human concerns, not in some quack professionalism with a Diploma as a magic key to the mysteries. From then on, to keep the faith, to keep open to the future, is to know the present as a commitment in depth, and to know the past where its spirit can still reach us. As to know your enemy is more usefully to know, and to seek, your friends. The enemy can usually be trusted to reveal himself (or is only too easy to manufacture).

A designer comes to recognize that his world is only fleetingly conjectural. Just as a student finds it hard to believe that anyone will actually give him a job – and there he is five years later, working and still managing to eat – so it is with all our legendary tomorrows: they offer their problems concretely enough, if we are *there*, and refuse to take no for an answer. It is also true that a designer interprets reality through the modalities of action; in the end, his work stands or falls by the intractably objective qualities of an outcome. Only in such good sense is he a philosopher: in making actual an experience of design, and thus constantly re-defining what the word (and the work) stand *for*. Yet because we occupy our differing roles only in virtue of our common predicament, which is our humanity, there is no end to the questions we must continue to ask.

Design is thus simultaneously a realm of values and a matter of engrossingly particular decisions, many of which are highly technical. There is a threshold up to which we can quantify, and this is often enough the task for a professional: less an equation of meaning than one of ordered evidence. Beyond this threshold, design is strictly a cultural option. It always has been. We humble ourselves, we sharpen our wits, and we offer, with all the resource of our persuasiveness, the fruits of a trained and straining consciousness; our moments of lucidity. Our concern is always 'the place of value in a world of facts', but there is no role waiting for us, there is merely the chance of making one out of the sheer courage of our perceptions. In the same way, if you want to link hands with the spirit of the modern movement, it won't come to meet you; you must go out and make it your own.

Part II: Notes on design procedure:
seeing a problem for what it is (and might become)

Does a respect for technique inhibit individuality? Try it this way. If I go to classes in seamanship and navigation, I shall listen humbly for every detail of practical advice. I shall not thank the instructor for telling me what boat I must sail, or where I ought to sail it to; though I may be interested in any suggestions. If he is a good instructor, I shall expect him to add two reservations to his teaching. First, that everything he says must be tested against my own experience. Second, that to get from A to B without danger to myself (and others) is one thing; to sail with art, and with satisfaction, is wholly another. That will be for me to explore.

Design work is like that, for the designer in a personal relation to his work. The work itself makes its own quite different demands.

7 Introduction

During the past 15 years several design schools have been testing the relevance of problem-solving methods to design procedure. In Europe substantial work has come from the Hochschule für Gestaltung at Ulm, Germany. In England there have been published accounts of such work in English schools, and theoretical papers, notably from L. Bruce Archer, John Christopher Jones, and more recently David Warren Piper. The work of Chris Alexander, and the continuing researches of Mrs Abercrombie, are slightly to one side of this fieldwork, but can be seen as contributions of quality. In America, Buckminster Fuller soldiers on, in pursuit of 'global strategies' that embrace an avowed problem-solving attitude, and – perhaps necessarily – a good deal else besides; and no doubt there will be a continuing issue of work from design schools throughout the world. Students will naturally be attracted to bold and speculative work in the field, but would be unwise to neglect those sources which derive from patient experiment.

The notes that follow are more to do with technique than method, and do *not* attempt a systematic account of what has been well said elsewhere. One difficulty is the mystification and dead language that sometimes surrounds an academic discussion of problem-solving, so that a sceptical student might dismiss the matter as a smoke screen thrown up to conceal a retreat from common-sense, or worse, a retreat from any ability to make simple decisions with confidence. The following pages are by no means free of jargon, but the commentary is intentionally an informal one. The techniques (and vehicles) discussed are indispensable to a designer anyway, even if often practised by rule-of-thumb. There is value in thinking about technique when its context is appreciated, and to some extent experienced. These notes should therefore be read as a commentary to experience, not as a substitute for it; though they may also usefully accompany the study of technique in a wider theoretical field. In the testing realm of practical work the very word 'problem' should not be accepted without questions asked.

The account of report-writing may seem fussily detailed if compared with the brusque coverage given to, say, 'sources of information'. The reason is that little has been said elsewhere to place report-writing in the general context of design work. It is difficult to write about 'questions' without becoming painfully abstract. In a design school, this is something to study in active cooperation with general-studies tutors. If students will merely foster their awareness that questions can be asked clumsily or with skill, coldly or with sympathetic insight, then a very large part of diagnostic technique will fall naturally into place.

At the cost of some repetition, the notes are written to be consulted separately in each section. In each case, the notes refer to design situations as they are; though within the limits set in the introduction to this book. How things might become is always an interesting problem.

8 When is an analytical approach necessary?

The case for analytical technique – of the simplest order – rests on the premise that you should find out what you are doing, or what you should be doing, before considering how to do it; and that this degree of foresight is normal to the role and situation of a designer. Embroidering the matter a little, a prescription will be a very hit-and-miss affair unless there has been a diagnosis, and a plan won't be much use unless it has been preceded by an accurate survey: in either case there are procedures that will help you to avoid wasted time and effort. So much is (nearly) self-evident; the argument about analytical method hinges on more ambitious claims sometimes urged for it – that given certain known procedures, a designer who employs them is conducted remorselessly to the 'optimum solution' that would otherwise have eluded him, or at best arrived by happy accident. The opposition says this is eyewash, problem-solvers are people who are afraid to use ordinary human judgement; prefer talking to action, and rarely make anything worth looking at anyway.

These are the extremes of misunderstanding. To see why (still in simple terms) it is necessary to think about the distinctions between different kinds of design job mentioned earlier in this book, and then to consult the description of his work in the discussion about designers and 'artists'. If you are making something yourself, you can freely set your own terms of reference. You can decide for instance, if it takes you that way, to make a chair the function of which is to question the notion of comfort, and which is therefore designedly uncomfortable to the point of actually collapsing when sat in (a 'sign for sitting' with question mark added). If you are commissioned to design a polypropylene chair for quantity production by a factory, you will be involved in an intricate investigation of the possible – in terms of marketing, cost, production techniques, user requirements – that will compel you to conduct research in a fairly methodical way. You will be acting as an agent in a social situation and it will help you to order and picture your allocation of effort. The form of the chair – assuming normal standards of comfort – will be to some extent preconditioned by what you will know or find out about the human body. Materials and form will be impinged upon by structural needs, perhaps revealed after prolonged testing of a prototype, and also by machining costs. The chair may be part of a 'set', a group of furniture intended to go together, in which case you will have to use a vocabulary that will fit all cases. This would be a product design job and product design involves a high degree of generality – the highest common factor of everybodys' needs.

A simple problem analysis would help you with this job, and would net for you, for instance, the required life of the chair as a genuine condition to first question and subsequently work with, but the scale and nature of the work would not require much more analysis than would come naturally to an intelligent person. Your real difficulty would be in satisfying the required conditions and designing what you felt to be a good chair out of them. 'Good' for you might mean an educated designer's appreciation of form nicely realized out of all the limiting conditions, one of which would be its acceptability to a less educated public. It will be clear that the margins of the designer's freedom in this case are tightly drawn, which is good reason for finding out how many of the restrictions are, in fact, mandatory, and how many may reflect habit and accident of thinking on the part of the client. This would be more obvious in cases (e.g. the design of some public area) that are more particular, where the client has asked for chairs but your analysis of the situation suggests that chairs are quite unnecessary. It is then the seating *situation* you will be examining – say, the need for seating places in some area of a public building, and your analysis may need you to reconsider the whole circulation pattern within the building, to find that if people are sitting down at all it is almost certainly in the wrong place and under the worst conditions for their comfort. Drawing on personal experience, I was once commissioned to re-furbish such a seating area which was unattractive, noisy, and getting badly knocked about (this was in a university). A careful analysis revealed that the unsatisfactory features of this area were merely a small symptom of major deficiencies in the total environment; in the end, the whole area was re-planned to include coffee bars, a large information area, private rooms for quiet seating, and even a complete alteration of staff offices, college entrances, and other facilities. Thus, through problem analysis, a small and apparently self-contained problem came to be seen at a different level altogether, one that involved not only something larger, but different in kind – the criteria appropriate became not only structural and aesthetic and practical, but sociological. The job had to be seen in terms of the conflicting needs of all the conflicting interests involved, which included staff-student relationships. Now if the resources to carry this out had not been available, or had been partially available, it would still have been desirable to place any temporary action in a broad framework of future possibilities – otherwise (whoever takes the decisions) the work would be stultified by false or insufficient terms of reference.

Again, the designer might recommend a completely different solution in which a series of experimental improvizations would be in constant change; he would not be designing any 'thing' as such, but recommending, and arguing for, a certain course of action that seemed to him would most fruitfully answer the problem (or, as he might argue, desirably continued to pose it). Or he might recommend an interim solution: to explore every possible way of using the spaces, thus accumulating evidence for a more structured solution that firmly organized the constants in the situation, leaving the variables to look after themselves.

In passing, it is worth remarking that a 'flexible' solution, theoretically at the free disposal of the user, can be both practically unsatisfactory and psychologically tyrannous, whilst (other things being equal) an apparently rigid and very definite organization of a space can bestow freedom.

To take further examples, there are some kinds of exhibition design, where to establish the central function of the exhibition considered as an effect, it will not be a matter of balancing out all the factors that must be satisfied, but if necessary sacrificing them to the demands of a single, powerful, and imaginative idea. Here it would seem that the designer either has a good idea or he hasn't. Yet to be sure that this is, in fact, the right approach, the designer will not only need the confidence of his idea, already as an asset in hand, but he will need carefully to think out his case for using it, against all the alternative approaches, and then persuade the client that he is right. In doing so, he may well find that this first idea must really give way to something more comprehensive in the total requirements of the job, even though he knows that an exhibition is an ephemeral occasion with a good deal of latitude in the design constraints.

Graphic designers very frequently begin with a false brief from their client which will need to be taken right back to its origins before design (in the ordinary sense) can begin. This applies less to bookwork than to the very wide field in which graphic designers are now employed. To keep to exhibitions, and assuming the job is large enough to involve a graphics specialist, he will be discussing with his design colleague not visual patternmaking but the nature and content of required information; whether, for instance, the purpose of the job would be better served by separating out one level of information from another, so that the exhibition visitor reads all about it at home from a pamphlet, while the sole function of the exhibition stand is to identify itself and to make sure that the pamphlet is effectively distributed. Such a decision, and the considerations that prompted it, would become integral with every other non-graphic decision in the design of the exhibition. This simple example points to the absurdity of educating designers in compartments, and of over-using words like '2-dimensional' – when was an exhibition ever a 2-dimensional experience, or a signing job in the grounds of a university, or a catalogue?

When design decisions have to do with a minimum of external constraints, and this might occur in any design field, not just textile design and ceramics where it is usually assumed to occur, or where the 'sense' of constraint occurs in an intimate dialogue between the designer and the nature of his materials, it is all the more necessary that the designer partakes fully and critically in the culture to which his work will form a contribution. This is not a matter of respectful attention to its substantive achievement, but a real involvement in the assumptions, questions, hopes, that the work subsumes, and – in terms of hardware – an understanding of what has been attempted in the field of object-relations. Without using too many special words here, the point may be illustrated as follows. A textile designer should be able to enter a modern building and have enough awareness of space and form relations to respond to it in a lively way. He will not have enough technical knowledge to estimate the restrictions, or the opportunities, that the architect has exploited, and therefore to form an accurate assessment of its achievement. But given the chance of an hour's conversation with the architect on the site, a fruitful exchange should at least be possible, and a part of it might require the mental shorthand that presupposes a common professional culture. To hammer home the point, a textile designer who conceives his work as self-expression and works solely alongside other textile designers, may make some nice things, but the work will be deprived of the strength and the insight of a broader cultural reference, a sense of weight and appropriateness in decision-making. Taking just historical examples, a textile designer who has never felt for the distinctions of outlook implied by, say, Ronchamp, the Barcelona Pavilion, and the Dymaxion house (a very indigestible mix), or has never contemplated the difference in living in a cube as distinct from the geometry of an inflated structure, is simply working as a badly educated specialist. If, on the other hand, the designer has this involvement, but then elects to focus his energies within a limited and specific field, then there is a good chance that his work will *refer outwards* and gain meaning thereby. His intuitions will have been instructed by a richer sense of the place he occupies. Here is the parallel with analytical procedure. In forming a hypothesis, the designer will not only have a 'picture' of the problem to offer and test it against – this is professional necessity – but he will also find that when a hypothesis springs into consciousness ('creating' a design) it is instructed, it 'fits' the conditions of the job rather than arbitrary conditions – and this is creative necessity. When something goes wrong, it will usually be traced back to the beginning, to a deficiency in the design concept which in turn derives from the acceptance of false premises, not the logical structure of the argument. This is why questions are important, and – essentially – the resourcefulness of attitude that prompts them.

One further example may serve to advance the discussion. A furniture designer is considering the design of a sideboard, and let us suppose that at this stage his terms of reference are open. He will know that the word 'sideboard' refers to a crystallizing-out, historically, of certain recognizable forms to provide for recognizable needs. He will question whether or not this is now a dead usage, better seen in fresh contexts which may suggest alternative forms. He will therefore examine the whole environmental situation to which the provision of a 'sideboard' has been, in the past, a part-answer. He may recall Rietveld's work on this problem, and what it had to say so presciently: 'here is a contingent equation, the elements and components of a sideboard reaching out and dissolving away into new meanings, new relationships with what surrounds them, a question implicit in an answer' (this is my own response, not Rietveld's: his work speaks for itself but also summons dialogue). As the task begins to set up its field of reference, he will find it helpful to use certain abstractions as a tool of analysis: perhaps the analytical categories surface, container, and support, qualified by secondary categories hold, connector, screen, mechanism, and so forth. He will relate this thinking to his analysis of the needs that have to be served – for instance, in the case of storage, what is being stored, what constants and variables occur, what levels and readiness of accessibility, does 'container' here imply just screening from dust and sight or is temperature control relevant, what are the frequencies and occasions of use.

This analysis will cross-reference, later, with a structural analysis and more material considerations. He has yet to consult the social and production constraints that determine the *possible*. If the job is undertaken for small-scale production, or in the one-off context or a furnishing contract, there will be far easier acceptance of a way-out solution than would be possible for mass-production.

Beginners who find this paragraph obscure, should move on to the following one. The hypothetical furniture designer is not departing from reality into abstraction; he is using his analysis in order to return to reality with a new understanding of his problem. He knows, or should know, that the difficult part of his work occurs, not in determining relationships, but in realizing them (materially). The design may exhibit – implicitly – an intellectual proposition, or an argument that can be 'read' and for which the premises may be clearly inferred; but this is unlikely to be more than one dimension of meaning in the job, and not necessarily the most essential one. What he is seeking analytically is a sense of priority in the job, the conditions of its meaning, and he may seek a 'diagram of correspondence' to help him – not to recognize and dodge his constraints, to get round them, but to take them up; to identify their potential and use this potential to positive advantage. The sideboard may well disappear as a recognizable entity and its functions disperse into new relationships. It is worth saying, however, that if this designer has some experience of organizing *information*, he will be greatly helped to re-order what is left in a way that is at once definite, intelligible, and open. 'Open-ness' is hard to define, though not difficult to recognize. It is the property of a solution (thing, place, message, situation) that enables life to flow through it, that makes its meaning accessible, indeed defined, by projections of reference that are strictly speaking outside the problem itself. Designers in some fields will have seen an object developing into a system, system into process, and process into information, in a way that disconcertingly suggests that their world is dissolving away around them. In making finite decisions in this context, they will be aware of a new order of reference in their thinking. This is the simplest consequence of a changing world-picture. Similarly, a highly particular design solution (including those with determinants from place and occasion) must now – if they are to live – not only exhibit a field of generic reference, but derive a gain in particularity from it. Anyone reading this book casually will be defeated in making sense of this, but I merely want to add that a designer attending to such considerations, will be actively involved with the balancing and weighting of categories and therefore the *kind* of analytical thinking that uses sequence, distinction, relation, hierarchy, and so forth, to organize information.

Thus the furniture designer's thinking can be joined with that of a graphic designer. Using a similar example (a designer reviewing the immediate history of work in his field) the furniture man will have pondered the subtle qualities that make Aalto's furniture so remarkable. He will know how little of this could proceed from any form of structural problem-analysis. But if he has also studied the beautifully articulated work of early Tschichold in typography, the extraordinarily subtle weighting of categories in conveying information, he will know that here are benefits of approach that can be transposed quite directly into his own field. I have used historical studies to simplify the discussion. In fact, in a livelier sense, any furniture designer who works in this way will approach a rectangle of paper in the same spirit as he would a rectangle of circulation space in a building; as a design opportunity.

Now there is nothing in all of this that could fail to come naturally to a designer thinking intelligently about his work. There are two conclusions: first, that analytical thinking will always tend to be interdisciplinary in its assumptions and its effects. The second is that this kind of thinking, together with problem-solving *procedures*, are inevitable to design work. But here it is necessary to distinguish procedure from method, and particularly from elaborately structured method.

For the range of work with which this book is concerned (an important reservation), it is always tempting to confuse logically or mechanically verifiable modes of thinking, with the informal, untidy, difficult, and intricate procedures that usually prompt a fruitful design hypothesis into being. Some of the best ideas occur on the backs of old envelopes. Computer-aided analytical techniques become necessary as the scale of the work increases, and as the conditions for its measurable efficiency become both critical and manifold. Where very complex constraints are involved, as for instance in planning the coordination of operations on a large building site, the job cannot be fumbled so far as an analysis of quantities is concerned. For any design work with expressive latitude it is still necessary to define that latitude. But in the work discussed here, analytical thinking has the secondary function of presenting the designer to the work responsively, and, for some people, a reliance on sequential method will paralyse rather than sharpen their responses.

The worst error is to take refuge in 'method' or 'process' at the expense of *any* practical commitment. Thus is a wilderness created by default, and argued for in retrospect by specious appeal to scientific method in the way the problem was approached. As Professor Medawar has reminded us, much scientific achievement has relied upon hunches which have *then* been scrupulously examined against existing schemata to see if they work. Not that such comparisons throw much useful light on the designer's role, which is to be available to tasks for which his experience is fitting, in full awareness that some things are better done by people for themselves.

A socially adaptive view of design practice takes the view that a designer accepts what he is given and makes the best of it by knowing how to weigh up his constraints shrewdly and logically. This is a perilously inadequate view of adaptation, (even), as can be seen in every area of society where new adaptations to reality are breaking through old and outworn assumptions. A designer must have his tools around him in the ordinary way of competence. A good joiner, who knows (as the saying is) seasoned wood from green, will have in his toolkit a few tools that are rarely used but irreplaceable when needed; he must know how to use them and when to pick them up. A designer's priority is openness of creative response, the capacity to mix the levels of his thinking, to ask productive questions; he must seek the personal conditions that help him to work in that way, and the social conditions that allow him to. This search will not prevent him getting on with the job at the limits of the possible, keeping all his tools sharp for the next chance to make himself useful.

9 How is design work conducted?

The step-by-step account that follows will place design procedure in the normal sequence of events, though one that has been both tidied-up and simplified. The job might be, say, a small department store or a fairly large exhibition; or if designers imagine the procedures applied to the fitting out of a large passenger ship interior, most specialist interests will be accommodated.

I have used the opportunity to give a running commentary. From such an account of happenings, it is possible to abstract out an analytical model of the design process, and thus to see what is irreducible over a wider range of problems. This would be unhelpful to the present discussion, but readers are referred to Bruce Archer's *The structure of design processes* for a systematic view in such terms.

1 Letter or telephone call from client or equivalent.
Arrange a meeting at his and your convenience.

2 Meet client, informal conversation; client airs his problems.
This is an unpredictable occasion, though the client will usually be anxious to convince himself that you are competent, experienced, personable, and able to look after his interests. Prepare accordingly. If you decide to take some work, a vast and shapeless portfolio may prove damaging by making the client feel that you are setting the pace, and in doing so, exploiting his own lack of technical expertise. Equally it is unwise to turn up with batteries of assistants, tape-recorders, and measuring tapes etc. It is most essential to *listen* properly, and to take notes; listening being that part of conversation that a designer must practise and get word-perfect. Briefs; see 15.

3 Visit site, meet other interested parties.
This is again an informal occasion – not a site survey – but an important one, because first impressions are important. Note them. The 'parties' may be the clients (plural) or the client's business associates. Multiple clients are usual in jobs of any medium/large scale, involving expenditure of (say) more than £5000. Take camera, but only use if the occasion seems suitable.

4 Exchange of letters (leading to contract or letter of contract)

Must be considered in formal terms; free of too many references to 'I', and not over-burdened by design jargon or unsuccessful attempts to sound like a business man (e.g. 'the favour of your letter' and 'may we assure you of our best attention at all times'). The best form usually is a direct and friendly letter, practical and reserved in tone, with facts or lists separated out as an accompanying sheet, such that the client can hand this round without his colleagues having to work through the more ephemeral or personal bits. At the beginning of a design job, it is often helpful to a client to have *a very simple* account of the design sequence that you usually follow, so the client knows what to expect and roughly when to expect it. Be careful, however, unless the job is one you have encountered before. A letter is a good vehicle for this simple information. You should advise a client that consultants may be necessary (engineer, quantity surveyor, etc.) and you should make him aware that formal consents may be involved, and that it may be necessary to consider his personal insurances, his obligations of tenancy, the rights of adjoining properties, planning and fire office consents, and (in the case of an exhibition) the exhibition regulations.

Obviously, letters continue to the sweet or bitter end of the design process.

5 Contract or letter of contract
   Legal documents, highly conventionalized. Should be separately studied as an aspect of 'professional practice', the subtleties of which are outside our present terms of reference.

6 Letters of record or enquiry to other interested parties
   Such letters will always be short and to the point – such letters are almost conventions and most of them go into the filing records of local authorities etc.

7 Site and premises survey, subsequently drawn up
   It will save endless work if this is done properly because the drawings must be very accurate descriptions of fact, accompanied by photographs, recordings if necessary, and extensive notes. Sometimes you will employ a surveyor to produce the measured drawing; you will still need your own observations for design purposes. Surveys will be considered in the detail required.

8 Questions
This is too important to be skimped; here, it should merely be noted that questions may be asked orally, in which case it is acceptable at this stage to use a tape recorder, or may be submitted rather more formally as a questionnaire. Questions will include requests for factual information available to the client (e.g. for a shop, very elaborate questions about his stock and sales policy). It is unwise to ask questions casually by letter or at random meetings. The client will usually respond well to carefully prepared questions; they are a measure of your concern for his interests. See later chapter on this aspect of design work.

9 Research
This will involve a consultation of all other parties or interests involved, particularly local authorities in the form of the District Surveyor, together with all necessary technical information (catalogues etc.) and anything else necessary to complete your reference data. A considerable task, and not one neatly confined to this stage of the work.

10 Sizing up the job
At this stage you should have enough information to work out an office programme with a clear allocation of personal responsibilities for the conduct of the job. This exercise will give you an intuitive idea of the possibilities latent in the design work (conditioned by time, money, and the nature of the problem) and just how far you can afford to develop them. This will affect your whole subsequent approach; your personal strategy.

11 Preliminary ideas
Design begins here in a formal sense, though you may by now have a design concept or a 'mental set' from your first meeting with the client or your first view of the site (this is a matter of experience). Mistrust mental sets until they have proved their relevance. This is the stage at which ideas are roughed out diagrammatically and working principles examined, tested, and agreed. Such ideas may take a mainly visual form (but will usually be diagrammatic in character) or, if you use a 'report' as a kind of developing discussion with yourself, may be verbal argument accompanied by concept diagrams (q.v.). It is difficult to generalize about the thinking that may go on at this stage.

An intricate problem may require some apparatus of formal analysis. This is the place for it.

12 Report
A report is not always necessary, but will usually repay the effort involved. See later chapter. The report will embody your proposals in principle and your reasons for suggested courses of action. The report may include diagrams and a plan layout, and may include catalogues or other references. The report should *not* commit you to what the job may finally look like. Desirably, your client should be left some leeway (as little as possible on any issues you know to be crucial) so that his own contribution can follow, or alternatively, you may wish to present several alternatives, leaving him to choose. It is fair to say that most clients find this very disturbing, but all clients appreciate a well-argued comparative survey of the possibilities, leading to the gist of your own recommendations.

13 Pause: client (board, directors, etc.) brood over report, which you may have sent by post or presented personally.

14 Meetings, conversations etc. to discuss the report and your client's reactions to it.

15 Brief
A 'brief' is a statement of your agreed terms of reference for developing a design to presentation stage. In a small or simple job the 'brief' will emerge much earlier in correspondence with your client. Here, the brief emerges from final agreement after the report has been fully discussed, with all modifications taken into account. It is highly desirable to set out your brief in a manner that obviates any kind of misunderstanding, but it is by no means an onerous task (not comparable in length or scope or detail with a report). Its purpose is to reassure the client that all his reservations have not been brushed aside, and to remind him that from now on work must be seriously under way without his interference. This is the last chance for the client to reject your approach to his problem (unless he finds your presentation drawings unacceptable).

16 Development
Here begins 'design' in the wholly conventional sense. You will

have a model from your survey; if not, make one now. According
to the nature of the job, you will proceed on the drawing board,
in the layout pad, and in your workshop if you have one
examining and developing ideas and testing them against your
problem analysis if you have one as such, or against the
reference data you have accumulated from your client and other
sources (or against your report if you used the report as an
analytical tool). You will discuss alternatives with trade
representatives, with your quantity surveyor if you have one,
with any relevant local authorities, and you will form a good
idea of the costs involved. All this will lead to a 'presentation'
of your ideas, or rather your intentions, to the client.

17 Presentation
This will normally involve a model, sample sheets, diagrams,
notes, sketches or 'perspectives', graphic layouts in whatever
form appropriate, etc. and will be personally 'presented' and
argued for by yourself to the client or more usually to his board
of directors. You may need an assistant, and, if it is a combined
graphic-construction job, both of you will be present. A design
presentation should go pleasantly enough *if* the work outlined
above has been carried out properly. A presentation should be
seen in definite terms (if it is inescapable) and any alternative
ideas should merely be supporting evidence for your proposals.

18 Modifications
Certain changes may be necessary in the light of your client's
response to the design presentation. Normally, such changes can
be agreed informally (covered by letters) in your own office, and
will not involve the effort of a further presentation.

19 Working drawings
The largest task now begins. Working drawings, or construction
drawings, are detailed accurate informative *instructions* on which
the work will be carried out. This is too complex to discuss here,
except to say that drawings for consent will normally be done
first, and the remaining bulk of the drawings must be carefully
planned for the intended contractor or sub-contractor or
manufacturer. Drawings may be submitted to a quantity surveyor
in order to receive a Bill of Quantities (a measured account of all
labour and material items involved in the job) or may join a
written specification and schedule in submission to competitive

tender. These are technical matters which need not worry you in this short summary. Discussions with contractors may involve many modifications as the work proceeds, but if the contractor is unknown (i.e. to be decided by tender) such modifications will ensue after the contract has been settled.

20 Tender
Here the picture must be simplified; readers are referred to Ron Green's *Architect's guide to running a job* and J. B. Creswell's *Honeywood File*. The latter is an amusing and instructive account of everything that can go wrong – somewhat Edwardian in flavour, but very readable.

Designers (in this case, as distinct from architects) will often work in an informal way with a contractor already well known; in which case much of this procedure loses its worrying aspects, but inescapably, however the job is done, a very large number of drawings must reach the contractor with sufficient notes or verbal description (as in a Specification) to leave no room for doubt.

21 Contractor begins work

22 Site supervision
This is an important part of design procedure. With the best will in the world, and a very good set of drawings, things will always go wrong on site and last-minute alterations will be essential. The designer must also make sure that deliveries are properly in hand. It is best to confine site instructions to the site foreman acting as a general co-ordinator. Visits will also be necessary to the joinery shop, to the printers or typesetters or indeed to any supplier or sub-contractor.

23 Payment
The designer will normally certify contractor's claims for submission to the client. In the case of shop-fitting or similar work, an agreed small proportion is held back for a six months' 'defect liability period'.

24 Inspection at completion stage, and subsequently after any defects are corrected.

25 Final settlement of finance and records. Photography. Filing and storing of records. Job over.

The above account is too tidy and logical, despite some rather alarming simplifications. In practice, constant new factors upset the brief, confusing or enriching it, and every kind of contingency will disturb a logical sequence. The design process is anyway a continuous interplay of creative thinking with reference data. However, in jobs of any substantial scale something like this apparent sequence is the only way to uncover the creative possibilities and to realize them. It will be seen that 'communication' becomes of first importance, with every new situation making its own quite distinct demands.

10 Communication for designers

Communication is a word that needs qualifying. If poetry achieves the most exact and comprehensive use of language as a mirror to experience, then one must say that communications-theory has rather special areas of experience in view, and what is mirrored there is not always of universal interest (or clarity). However, Colin Cherry's book (see references) is a good plain man's starting point.

Here, we shall discuss the ways and means of sending and receiving and getting information relevant to a purpose in hand, that purpose being the whole conduct of the design process as it affects a designer, his colleagues, the clients for whom he works, the contractor or manufacturer, the user or 'public', and all other parties or interests involved. This is a laboured definition, but it must be clear that here is purposeful communication used as a means – not an end – in situations of which the outlines are fairly well known. Graphic designers can find this confusing, because their 'end' may be to find ways of transmitting a given message effectively. Thus they are using, in the design process, two lines of communication that may either cross or reinforce each other. Other students will be confused by the general tendency of design school drawings, reports, etc. to become in themselves an end rather than a means. Students will also have an optimistic impression of the time normally available for design decisions.

The most cursory study of the events that configure the 'design process' will reveal that effective communication is of their nature. Very many forms and procedures are involved whereby clients are questioned, research is carried out, information is sifted and stored, matters are discussed and intentions aired or confirmed, designs are initiated and developed, and action made appropriate by laborious processes of instruction.

An experienced designer will call upon his experience; but will still find it useful to ask himself certain questions as he prepares a drawing, writes a letter, or considers the scale and purpose of a model. The following notes place such questions in their context and suggest something of their nature; though until the evidence has been experienced or at least examined in some detail, it is wise to approach any procedural suggestions with some caution. It is not a formula that is required; merely, at this stage, a few guiding considerations.

Looking to the 'job sequence' as evidence, it will be clear that the designer 'communicates' with himself – he exteriorizes his own thinking in drawings or, say, in using a report to think with – and with his colleagues in ways that may be highly personal, yet will have to assume recognizable conventions to the extent that some external response, or feedback, is called for. As the work moves away from him so must his chosen means become precise, until, finally, his instructions to a contractor must be entirely free of ambiguity. Yet a survey drawing, done at the beginning of a job, must be equally matter-of-fact. Again, there is no hard-and-fast distinction between words and drawings, and even while setting out a survey drawing a designer will be thinking in words about the drawing itself and the physical situation of which the drawing is an abstracted analogue. Is it possible, then, to do more than feel one's way into an appropriate course of action?

To answer this question, it is necessary to see what is typically achieved by communication methods at every stage of the design process, and then to decide if there are distinctions of choice that could be critical to its effectiveness.

Design methodology can usefully borrow from medicine a distinction between diagnostic and prescriptive procedures, though only a design consultant operates in so Olympian a spirit; the ordinary designer goes on to administer the medicine and thus combines the functions of nurse, chemist, and almoner in ensuring his patient's full recovery. Anyway, the idea of diagnosis and prescription is helpful. Broadly speaking, diagnostic work involves the classical apparatus of problem analysis, in greater or less degree according to the complexity of the job, and to the extent that the job does, in fact, constitute a 'problem'. (It is a mistake to suppose that all design jobs are best understood as problem-solving.) In the context we have used as an illustration, there will be three stages of diagnosis:

1 finding out: to observe, measure, assess, question and record – to get the facts and sense impressions which enable

2 sorting: to compare, distinguish, relate, and order the phenomena with which the designer is confronted, to 'un-scramble' the mix and to find sets or categories which helpfully accomplish this

3 interpreting: to evaluate the situation thus exposed, to examine its potential in terms of means, effort, scale, kind, and so forth; to

adduce the principles and the alternatives on which a prescriptive solution might be based.

Anyone who credits these procedures with 'objectivity' must be disenchanted. Human judgement will colour and preselect every seemingly objective assessment other than measurable fact reducible to number (even there strange things can happen) and problem analysis would be a dull and fruitless affair if it were not so. All the more reason, then, for scruple, and for technique. Returning to the analogy with medicine, signs are more easily detected than symptoms. A designer who approaches his task with the singlemindedness of an adding machine must expect a barren diagnosis and often a misleading one. Imagination and empathy are as necessary as keen observation; to use a different vocabulary, the designer has to put himself out to meet the new situation with all his faculties as a human being, not merely shelter behind the privilege of his role-function (shielded by what is expected of him).

As has been implied, problem analysis will be more or less useful according to the sensitivity with which it is programmed, but it will not lead to an optimum solution so much as an awareness of the conditions for a relevant solution – in other words again, if badly used, it becomes a device of cyclic definition for second-rate value judgements. In some cases a diagrammatic answer will cover the facts, with little to spare, and this is not an attainment to be despised because at least the brief has been competently respected. However, prescriptive work is often too particular to a given problem to stand much in the way of generalities. What is certain, is the crucial effect of properly used communication at every stage. Not only can misdirected effort be minimized, but bridges can be formed between latent possibilities and a designer's personal vocabulary or resourcefulness.

The following short statement puts the matter in a nutshell and suggests a simple way of evaluating any communication opportunity involving sender and receiver.

Communication in the design process:

1   is purposive (leads to or qualifies action in a design context)

2   is not random, speculative, a work of art, or 'self-contained'

3   does usually presuppose a known context, sender, and receiver (or class of receivers)

4   should effect its purpose(s) economically

5   must therefore be *clear, full, simple, direct, necessary, and acceptable*

6   unless your purpose is better effected otherwise
    NB for a document to be acceptable it must employ the right tone of voice for receiver and purpose (this applies as much to layout and materials as to choice of words and syntax)
    for a drawing to be acceptable, its purpose must be estimated with precision, and the drawing must use recognizable 'conventions' to optimum effect

7   ask yourself:
    why this document or drawing? (or model, or . . .)
    for whom and to whom and who from?
    in what context?
    for what purpose, with what intention, with what (precise) aim?
    or which of several (possibly related) purposes?

8   then ask yourself:
    when it should be done and when conveyed
    how it should be done and conveyed (less question than answer)
    NB why – who – what – which – when – HOW

9   consider context, relevance, completeness, sequence, tone, consistency, handling, filing, storing, copying, numbering, and . . ?

10 by these means make quite clear to yourself the *purpose, nature*, and *occasion* of what you are doing which will enable you to select the best means available,
    in a word, the *form* of your communication,
    (derived from conscious choice of mode, vehicle, media, agents).
    NB for the designer, every communication problem is a design opportunity, and good design may be held to imply not an identity of form and meaning but a close isomorphic relation between them. Yes, no, maybe . . . ?

11 the case for navigation by precept being that these are the shallows of communication (where it is easiest to go aground)

12 This requires the special use of four words (replaceable).

mode: way in which a thing is done, class of thinking or intention exhibited – e.g. might be persuasive or informative, analytical or descriptive, interpretive, literal, final, provisional, open or closed as regards conventions, etc.

vehicle: known form or apparatus of communication, recognizable by conventions – e.g. might be a perspective drawing, a diagram, a report, a questionnaire, etc.

media: physical means employed – e.g. might be card, cartridge paper, graph paper, wire, perspex, etc.

agent: instrumental means, tool or process – e.g. might be pen, typewriter, Xerox, tape, duplicating machine, etc.

13 Because the purpose of any given communication will not be terminal, but part of a network of interchanges of extremely various nature, the occasion or context must never be left out of account; it is not sufficient to think of a simple loop diagram with feedback. Information circuitry is as useful, and as misleading, as conceptual models of the design process in other terms: the dangers are inhibited response, a false sense of 'objectivity' in procedures, rigidly estimated criteria, and a diminished sense of reality in a designer's informal relation to his work. Although such hazards can be described with some subtlety in cybernetic terms, there is no evidence that an acquaintance with information theory helps to bring home the bacon, in the range of opportunities discussed in this book – and some evidence that reliance upon it will produce a bag of bones. Students must unravel this matter for themselves.

14 If any student finds the suggestions in this chapter uncongenial, or mystifying, he should think of alternatives – providing he is aware of the difficulties that are pointed to.

15 It is rash to assume that the final relieved handshake is 'terminal' to a job – as newcomers will soon discover.

11 Drawings and models

It is very important to distinguish one kind of drawing or model from another – by never failing to consider its purpose, occasion, and recipient, and therefore its nature. Confusion would be avoided, and tedium minimized, if draughtsmanship were always studied in the general context of communications theory and practice; but this does not frequently happen.

Fast and skilful technical drawing as carried out in a design office is just a matter of long practice: students should not be discouraged by their first indistinct attempts. I am speaking solely of competence in technical drawing; not of satisfaction. There appear to be no short cuts to this technical facility but common sense and a clear definition of a drawing's purpose will give a head start. Free-hand drawing is something of a personal gift, valuable to a designer but not as essential to every kind of work as is usually thought – a designer can design well and not be able to draw in this way at all; conversely, a designer who draws marvellously may be mistaking his vocation as an illustrator or painter. Much design drawing – of an informal kind – is no more technically 'gifted' than mental sums transferred on to paper. Designers think and talk with sketches and diagrams, sections, full-size details; a fluent and personal use of diagrammatic technique is certainly necessary. Such work can be better seen as analytical mark-making than as analytical drawing in the classical sense. Observational drawing helps focus observation, and has all the well-attested and richer benefits from a coordination of hand and eye, but the transference of such gains into design procedure is an indirect one – as would be from comparable activity of other kinds.

Nothing should be taken for granted in the way of materials and instruments. The qualitative differences must be explored, compared, contrasted; if only because much of the pleasure of drawing (or modelmaking) comes from a subtle rightness in the relation of tool or instrument to material. Anyone who uses blunt tools, or confuses a Pelikan Graphos with a clutch pencil, or who treats paper as a neutral non-material, will miss the benefit of good technique as a feedback into design thinking. These are fine-tolerance perceptions carried out at the finger tips, but awareness of them can help the tone of design responses in a very encouraging way.

Drawing practice and modelmaking will occupy a large part of a designer's time, and in a design school, even more. There will always be formal instruction in schools, but the following simple distinctions will help to place these skills in a communication context:

Diagrams are abstract, partial, energetic, concerned to establish or convey ideas and values directly, thus having an analytical or interpretive purpose. Usually have open conventions (excepting graphs and mathematical conventions), may be imprecise, or may be examining exact quantities, usually have diagnostic function.

Illustrations are depictive, present appearances from which inferences may be drawn, are often atmospheric in nature and persuasive in purpose, have closed conventions. Usually have prescriptive function; better for presenting conclusions than determining them.

Surveys are records of measured and verifiable fact reduced to quantities, though survey drawings may be accompanied by interpretive notes. Closed conventions. Diagnostic function.

Working drawings are strictly purposeful and are instructions. Use rigid but propulsive conventions (i.e. lead to required action). Many types according to purpose, occasion, and recipient. Prescriptive function.

Signs, words, numbers, codes, etc. – clarify, supplement, regulate.

Warning: let such distinctions only be your starting point; a designer thinking out a job will go backwards and forwards from one kind of drawing to another; testing and exploring. Thus, as was said of communications in general, there will be a balance between subjective freedom and objective scruple; as the job develops 'the chosen means become precise'.

12 Asking questions

The notes are confined to enquiry in the particulars of design practice. More general and more searching questions are professionally relevant – to fail to ask them is to invite creative anaemia – but such questioning is ill-served by precept. Though perhaps it helps to jump over mole-hills before climbing mountains.

As we have seen, most design problems will be presented to you in ways that may be diffuse, ill-defined, or actually misleading; yet, as we have also seen, a problem cannot be resolved in any satisfactory sense before its nature has been determined. According to the 'problem', such discovery may be instantaneous, or the problem itself simply invented by the designer in an extremely open situation; in either of these cases it is unhelpful to use the word, or to conceive of design as problem-solving: a different order of resourcefulness will be required. However, keeping to the work which has been used as an illustration, the designer will be asking himself and others a continuous stream of questions, some of which may need to be formally stated or written down.

The reasons for this:

1 To gain an awareness of the problem that is sufficiently objective, i.e. takes full account of relevant facts, relevant interests, relevant possibilities, relevant limitations.
2 To focus and clarify your own response to the job and to exclude (as far as possible) irrelevant responses – those that might belong to different jobs still strong in memory, that might be a simple projection of your own interests, or that might pre-judge certain issues before you have investigated them.
3 To enable you to sense out the feel, weight, context of the job and relate yourself productively to its potential (to turn yourself toward it).
4 To enable you to construct a satisfactory working brief, i.e. clear and agreed terms of reference.

General considerations:

1 The extent to which formal questioning is necessary will vary from job to job, but has nothing to do with the *apparent* scale or simplicity of the problem you encounter – questioning may

extend or alter the possibilities beyond recognition.
2 It should be recognized that diagnostic work has a creative component, a kind of dialogue between you and the client or between you and the total situation you are examining; thus its instrumental success very much depends on your own attitude and approach. To ask the right questions is not to carry out a tiresome preliminary to design; it is already design, and it may require considerable imaginative effort on your part.
In the answer to one question is the genesis of the next.
3 Questions served up 'cold' will harvest facts, but may leave untouched all manner of subjective or non-measurable factors which may be crucial to the understanding of a problem. Most (not all) problems will involve a client; most will involve human relationships, and therefore tact and an effort of identification with other people's viewpoints.
4 Your intention in asking questions must be reflected both in their structure and in their tone; categories of question useful to you may not be useful to the client or invoke helpful associations in his mind – only later, as abstract categories assume concrete reference, will you be talking on convergent terms with your client.
5 (Tone: 'the speaker chooses or arranges his words differently as his audience varies, in automatic or deliberate recognition of his relation to them. The tone of his utterance reflects his awareness of this relation, his sense of how he stands toward those he is addressing' – I. A. Richards). A suitable tone does not always imply informality, or prevent you from arranging your questions into categories, but may well require a careful choice of terms in describing them or introducing them.
6 The designer is acting as an agent in a process which (viewed in retrospect) may appear to have discovered itself over an extended period of time; questioning takes place across the flux of decision-making in a variety of situations. Early on, personal observation will involve self-questioning.
7 There is often value in mixing questions by free association; but concealed or subsequent categorizing will in itself provide association, and categories relevant to questions will usually be relevant to the answers.
8 Many questions are best put in a form that suggests typical answers, or alternatively put as assumptions that can be agreed with or corrected. Such assumptions need framing with some care, or you may cut out their possible suggestiveness and thus

'short circuit' back to where you started. Stated assumptions work best at the extremes – either as a mild confirmation of something fairly obvious, or as something so outrageous that a lively response is inevitable.

9 It is a mistake to rely on formal means to establish informal truths. Facts are best investigated formally; opinions or attitudes informally. A client may shy away from an attitude imputed to him by the way you have put a question in writing, whereas face to face things would have been different.

10 In formulating written questions, remember it is difficult for someone to withdraw a direct answer to which he has committed himself, unless you can show reason that he was misinformed. Commitment may be useful on matters of fact, but on matters of opinion may merely express prejudice. Taking this further, if you are working with someone toward some common end, you must allow their prejudice as much breathing space as your own; but don't carelessly drive them into positions from which they can't retract without loss of face, and which may come to impinge on your common activity quite needlessly.

11 Of course facts and opinions will be inextricably mixed in your client's experience of his problem; hence the delicacy of your diagnostic technique.

12 As a rider to that, be careful not to confuse diagnosis with cure. Questions seek answers; not a euphoria of perpetual doubt.

13 Loosely and vaguely-put written questions are quite useless. A 'portmanteau' question will receive a portmanteau answer, and also diminish your client's confidence ('what is he getting at?'). Questions are best put singly and profusely – unless you have a particular purpose in mind.

14 Questions and answers qualify and reinforce – but do not substitute – judgement and decision.

15 Consider carefully when and when not to use tapes, notebooks, etc.

Oral questioning at informal meetings:

1 Above all, remember that to question is not to interrogate.
2 Mental alertness is essential (not so obvious).
3 Leading questions may be fruitful as guides to intention or attitude in those present; questions can be put in this way at such meetings that could not be asked in more formal circumstances (or in writing).

4 Do not take your answers too literally – unless they refer to matters of fact (which can always be checked) it is necessary to take conversational answers or assertions at less than their face value.
5 It is difficult to establish needs by direct questioning, but relatively easy to establish desires, preferences, prejudice, and opinion based on specific experiences. Motivation is far more elusive.
6 Do not allow the nature or sequence of questions to get on to 'tramlines' – closed patterns of thinking – which may give you a false run of evidence.
7 Avoid giving an impression (which should anyway be unwarranted) that you have made up your mind and are merely asking questions as a matter of form.
8 Make some allowance for positive and negative appearing in a misleading juxtaposition: e.g. if challenged, a person with authoritarian habits may stress his anti-authoritarian ideas or desires. Most people have inbuilt compensations for exaggerated personality traits and they are delighted to be able to exercise them.
9 It is off-putting to ask questions conversationally from your notebook.
10 A useful extension of questioning technique is the meeting at which all parties concerned air their views at random, with you keeping the thing going as a sort of group therapist. The purpose of such a meeting should be understood by all concerned. Avoid the meeting becoming a collective instructional session leaving no subsequent room for manoeuvre. Such meetings can be timesavers, particularly where many people are involved, but they should be followed up by more scrupulous enquiry, whether or not such enquiry starts from inferences drawn from the meeting.

NB. In all this the designer is meeting a situation, not attempting to master it. The only value of conscious technique here is in making the designer more useful, because more accurately responsive, to the situation he is serving, and perhaps a bit less liable to take his own unquestioning assumptions for granted.

13   Seeking information

Little need be said, because all that is needed is initiative, scepticism, and a little method. Add energy; and stir with a natural inquisitiveness. Awareness of the telephone will help, knowledge of libraries and classification systems is necessary. Random or accidental information is often as useful – and from mixed sources. Backyard sources especially. It is easy to think that a mass of information is useful, when in fact all it may do is clog the imagination, suggest complicated ways of doing simple things, and discourage improvization. The important thing is to have a good knowledge of sources – where to go for what, when; and some overall picture of availability to back it up. It is therefore a good idea to form your own reference file. Rather than interfere with this task of survey – it is that – the following is just the beginning of a check list. Every item should have qualification – for instance, there are many kinds of library, some specialized. Do you know all the services available, the classification system, how to get scarce books or copies of documents? If not, a few weeks' work on information sources would be usefully spent . . .

Libraries
Bookshops
Standard reference books
Government publications
Catalogues (manufacturers and services)
Standards institutes and trade institutes
Design magazines and trade magazines
Permanent exhibitions
Annual exhibitions
Design and building centres
Trade associations
Shops (general)
Consumers' guides and testing agencies
Trade representatives
Officials from public bodies
Pocketbooks
Warehouses, yards
Magazine classified information
Annual trade surveys
Research institutes and papers
Factories and tool shops
Materials' shops

14 Reports and report writing

A report is a written statement of fact and opinion, addressed to clients or to colleagues (or to oneself). A report may be retrospective, and therefore a terminal objective in some series of happenings, but usually its purpose is seen as a more active one: to support, explain, or gain agreement to a suggested course of action.

Designers have good reasons for employing the report as a familiar and often indispensable communications vehicle. Used properly, a report will more than repay the labour of its preparation. A report can describe, interpret, and analyse a situation with sensitivity and suggestiveness, and in language that a client will at once understand. For clients, there are no transposition problems to inhibit full discussion with business colleagues. In the early stages of a job, a client may be wary of a designer's drawings, however seductively presented: the medium is not his own, and may conceal all manner of implied decisions which he can only take on trust. Sometimes the client will discharge his unease by demanding entirely unnecessary alterations, or he may even reject the first design out of hand, until he has been reassured by a set of drawn-out alternatives. This is a very hit-and-miss way of going on. A preliminary report gives confidence, because here is evidence that the problems have been examined with professional thoroughness. There is also the possibility of active intervention on his part *before* the design reaches the drawing board. A client who enjoys a real sense of participation in all the thinking that surrounds the design process may be ready to accept a radical and (for him) unfamiliar design solution, simply because he has understood the genesis of the solution in the problem itself. Thus may a very conservative client find himself embarking upon inconceivable adventures, and enjoying them.

These are obvious benefits to a designer's work, but reports can do far more than this primitive midwifery. In work of complex nature, reports may have the key role of a more subtle, creative, and constituent process. The writing of a report will order (and externalize) a designer's own thinking, and will enable him to call in question his own working assumptions – before it is too late. The designer will not only avoid misunderstandings with his client – a negative gain – but may also attain a new understanding of the task, both from the discussion that such a report will inevitably precipitate, and from the effort involved in arguing out a problem analysis. Normally a report is neither a brief nor an analysis as such, but rather a distillation from analytical thinking; expressed in terms which will further a diagnosis into relevant action. A brief must always be redefined when feasibility studies are complete, or when the problem can be freshly seen after due thought and examination by the designer. A report may include a suggested redefinition, or may merely point to its necessity. Finally, it must be realized that a large-scale job is a cooperative venture. It is a courtesy to give all concerned a document that makes plain the reasons for a design and their own part in it (this applies to design colleagues and consultants, but also to contractors).

It is sometimes said that all this is far better done over a glass of sherry at the club (designers are assumed to feel at home in all possible worlds) or perhaps during the main course of one of those interminable businessmen's lunches. In practice, a designer will have any number of informal meetings with his clients, at any one of which the burden of a report may be freely discussed. A report is a formal document that can be taken home and considered at leisure, or may be passed for comment to business associates, and to others whose interests may be involved. A report is also, indubitably, a record (though not a legally binding one). Reports have their own usefulness – they are not a substitute for conversations, contracts, letters, questionnaires, drawings, or other vehicles of communication.

Reports should be reasonably concise and readable (obviously), but beginners sometimes forget that style (meaning here, tone of voice) is equally important. Those who lack experience of writing should use short words and short sentences, making quite sure that the structure of the report has within it a clear and logical sequence. More confident writers will tend to adjust the tone of their writing to their purpose and to their awareness of the reader. At all costs avoid bombast, rhetoric, excessive technicality, and those knowing turns of phrase that imply ignorance on the part of the reader. A report so written may have unfortunate consequences.

The following notes summarize the active functions of a report, and the more obvious factors that should be kept in mind. There are less stringent requirements for the kind of report that merely supplements a set of detailed proposals (e.g. presentation drawings).

Reports:
1. must be written in full awareness of their purpose and occasion and the receiver (the nature, purpose, and occasion of a report should always be stated at the beginning, however informally)
2. require an imaginative awareness of the person(s) addressed, and thus an appropriate *tone* of address
3. should communicate in a direct, sequential, and consistent manner (derived from 1 and 2)
4. have no 'correct' structure, merely an appropriate one
5. will generally follow the logical sequence of given-required-proposed, though not necessarily a *formal* structure
6. should normally keep clear of design jargon and private language, using the word 'I' with discretion
7. should argue syntactically, adding questions or known facts or quantities in tabular form (preferably separated as a supplement to the report)
8. may include sketches, diagrams, graphs, photographs, etc. either as a supplement or directly related to the text.
9. if lengthy, should use short paragraphs which can be numbered to facilitate reference back
10. if lengthy, should also begin with a summary of conclusions but not, desirably, a summary of recommendations as such (better to let the argument require them)
11. may be read (and handled) by several people; to advantage, several (numbered) copies may be offered
12. must be complete and self-explanatory (the writer may not be present to make good any omissions)
13. should provide a professional interpretation of the problem that concerns your client
14. should thus establish for the client the full implications of the brief he has given the designer
15. should relate such implications to the 'constraints' in the problem (e.g. time, money, space, legal restrictions, etc. etc.)
16. should thus provide both an insight into the client's needs and the intentions behind the brief, and an estimate of the true possibilities latent in the brief
17. should thus clarify for designer and client the margin of free choice available
18. should certainly make clear the *context* of present decisions or recommended action
19. unless concerned with specific decisions, will be concerned to adduce relevant *principles* which such action must respect

20 most important of all, should gain a sense of *priority* both in the needs and potential of a situation
21 should seek a client's agreement to a designer's working assumptions, and any forseeable issue of principle that may bias the design approach
22 should recommend and define subsequent action (how, when, what)
23 may anticipate (require, provide for) a positive response from the client or persons addressed
24 may lead to a fresh and more explicit definition of the design brief
25 should be a creative task in the design process (constituent, not accessory)
26 should thus be directly helpful to the designer, and may even in some cases be written to himself or to colleagues concerned with the job
27 should be regarded as a design opportunity – construction, layout, materials, should not be overworked in relation to the content of the report, but should look and feel *right* in terms of 1–4

Appendix I: Notes on some of the references

It is difficult to decide in retrospect what books have been useful in writing this one; much of my own stimulus comes from other sources, and most of the comment, from personal experience of teaching and design.

The quote from le Corbusier is taken from 'If I had to teach you architecture' in an issue of *Focus* magazine (OP). The Conrad reference is from *Victory;* it is Heyst speaking. Tom Woolley is quoted from an issue of *Anarchy* magazine on architects and people (no 97); elsewhere in available back copies will be found Dr Grey Walter's 'Development and Significance of Cybernetics' (no 25) and subsequent discussion on the cybernetics of self-organizing systems by John McEwan (no 31) which may qualify the first three quotations in an interesting way. (Freedom Press, 84a Whitechapel High Street, London E1). F.R.Leavis' *The common pursuit*, I.A.Richard's *Practical Criticism*, and Ezra Pound's *ABC of Reading* are useful designer's books from another field. There are suggestive books on education by Paul Ritter, A.S.Neill, Paul Goodman, and Gerald Collier, among others. Recent student action is discussed in the Penguin books *The Hornsey Affair* (for evidence from the other side see *Report of the Select Committee*, HMSO) and *Student Power* edited by Alexander Cockburn and Robin Blackburn, and in the Panther book *The Beginning of the End*, by Tom Nairn and Angelo Quattrochi. Colin Cherry's *On Human Communication* is probably the clearest introduction to that field. I have found A.J.Ayer's books helpful (e.g. *Language, Truth and Logic*), together with books like Jane Abercrombie's *Anatomy of Judgement* and Susan Stebbings' *Thinking to Some Purpose*. Other books that spring to mind are Ruth Benedict's *Patterns of Culture*, Alex Comfort's *Nature of Human Nature*, Morris' *Signs, language and behaviour,* Gibson's books on perception theory, and books on sociology and pyschology (including Gestalt) that students can perfectly well discover for themselves. The word 'freedom' makes several tantalizing appearances: the view of freedom implicit here is examined in Fromm's *Fear of Freedom*, Martin Buber's *Between Man and Man*, Herbert Read's *Anarchy and Order*, Vernon Richard's *Malatesta*, and similar sources.

This is an unbalanced and perhaps rather curious set of references, but may serve to untidy the usual 'set-list' which is certainly thin stuff for any active minded reader. For those who don't read much,

I would add that a high score in personal literacy has no obvious or proven bearing on design ability (or intelligence), so don't feel undermined.

There have been a few books that examine design problems in some reasonable extent (Roth's *New Architecture* was such) and a few that compare notes on working attitude or objectives – *CIAM in Otterloo* was a good example of that. Ordinary picture books are less helpful than students commonly suppose. Little is usually shown of the origin, context, quality, performance, or even the form, of any design solution (it is so tempting to take nice angle shots). Students who come to rely on magazine culture can acquire a mental top-dressing of irrelevant imagery, which is hardly helpful in trying to design from first principles.

The mainstay of design reading is in the field of technical books or research papers, and also trade catalogues and catalogues of tools and equipment in the various fields they are concerned with. This is happy grazing pasture for most designers, and productive it can be. There are just a few out-of-print books about design that have their own object-quality, and come to the hand with the talismanic properties of a good ironmongery catalogue (*Der Stuhl*, the original edition of *Typographische Gestaltung*, Martin's *Flat Book* are examples) and designers can sometimes pick these up in the second-hand shops.

Synoptic books – those which place design in historical or theoretical context – are sometimes read more from duty than love, but the good ones (e.g. Mumford and Giedion) more than repay the effort. Students acquainting themselves with the history of the modern movement are recommended to go straight to original sources. Gropius, Rietveld, Corbusier, Moholy-Nagy, and others, have all written far better than the critics who have followed along with an assessment of their work. (So have Marx, Freud, and Einstein.) The spirit is there, and unless one is merely after the facts of history, that is what matters.

Summary:

What books are useful to a designer? Answer: any book that makes him want to design, and that helps him to design well. If this is fair, then a booklist is not only a personal matter, but it will include books with no ostensible bearing on 'design' at all. Such books are propulsive toward certain attitudes of mind useful to designers. Such books may be considered as 'tools' or propulsive agents, unlike those excellent books that leave one heavy and replete as after a good meal. The same goes for other sources – e.g. music, philosophy, mathematics. Or consider exhibitions. What design student has not been more stimulated by a Mechanical Handling exhibition than by a Furniture exhibition (or similar)? It remains true that there are a few non-technical books that almost every designer will have on his shelf, e.g.:

Corbusier *Towards a New Architecture*
Rasmussen *Experiencing architecture*
Gropius *New Architecture and the Bauhaus*
Gropius/Bayer *Bauhaus* 1919–28
Moholy-Nagy *The New Vision*
Read *Art and Industry*
Mumford *Art and Technics*
Giedion *Space, Time, and Architecture* et seq.
Pevsner *Pioneers of the Modern Movement*
Lethaby *Architecture*
Alexander *Notes on the Synthesis of Form*
Archer *Structure of the Design Process*
Banham *Theory of Design in the First Machine Age*

Appendix II: Suggestions for beginners

As a counter to the rather condensed argument in some parts of this book, what follows is a random mix of questions and advice written out very swiftly. Like a bran-tub, there should be something in it for everyone, but necessarily on a take-it-or-leave-it basis. I hope at best to provoke some suggestive thinking. Clever or disenchanted students can concern themselves with what is left out. Warning: test and evaluate for yourselves.

1. In first year remember you will reject everything before you complete it, because your values will change so fast. Can you turn this fact to positive account in the way you design? How?

2. Why not multiple solutions in principle to a problem if unable to develop one in particular?

3. Conceive the visible outcome of your work – including notes – as a totality; try to present a running sequential account of your thinking from beginning to end (using diagrams) for someone who doesn't know you or the problem.

4. Question every brief and state your own assumptions, briefly, so you know where you are.

5. In many studio projects there is an academic reality – work for you and your tutor – and an 'as-if' reality – work on the job itself. Should you separate them. How?

6. If you are miserably dissatisfied with your work on a job, make your answer to it a detailed self-criticism (a graphic project). Is this academic reality?

7. Out of every job that seems an indistinct mess, try to rescue one small part that is clear, simple, definite, and very well made or done.

8. Use colours freely in a layout pad; if you are beginning, don't sit facing tracts of empty white paper. Study the exact and detailed nature of all given factors in a job, work outwards from them.

9   Every student understandably begins by striving after originality. After five years work he is delighted if he can attend to a simple job with scruple and insight (unless he is about to launch a successful career as a carrion artist, see quote). It helps, at least, to know that.

10 If you think someone in your group has a better design concept for a job than you have, why not accept and develop it in your own way? The end-result will be very different, and a comparison valuable. You may have the best approach to the next job. Work toward objective standards.

11 Attitude: if you climb on top of a job, trying to master it, the work will suffocate. *Let it take you, play with it, search for its own life.*

12 In the way of samples or materials or catalogues collect everything you like or that for some unknown reason holds your attention; not what you ought to like. Information ought to keep pace with your ability to use it.

13 Don't be conned into thinking that only new materials or processes are worth investigating. Every material available is strictly contemporary.

14 Before deciding that Corbusier or Frank Lloyd-Wright are thankfully the last of the monumental masons, hitch across Europe or America and see for yourself. On the way, get right off the beaten track and study all the most humble of human artefacts.

15 If you must flip through photographs of other people's work, try this: write a short critical commentary on just one photograph, compare notes with someone. You may be surprised what the eye and intelligence gain from *focus*.

16 If the world is crowded with inessential rubbish, is there a case for seeing what you can do with the cheapest most simple and most ordinary materials?

Have you considered the fine distinctions that make human faces recognizable? – the elements are as much alike, and related, as you would get from a problem analysis. So are you just beginning? How do you see a face? Compare gasometers, pylons, cooling towers, street lighting. What is it about cluster high-lighting that makes the rest look sentimental? If you go into a furniture shop, examine the backs and insides and know before leaving how it is (or could be) made. Watch a road as a shallow kinetic relief, tread your way. Street furniture, what makes a good bollard? subtle geometry, spacing, linking, figure-to-ground relation, context, signal, available inference, finish, material? Look at a car and de-gloss it in your mind's eye, what of its form is shine? What about reflection in a glass building? Compare the cut-off of high-rise buildings, vertically and horizontally. Are some visually better than others, if so, why? Is a matchbox fine-tolerance cabinetmaking? Why does it work so well? When should a drawer be a tray? In typography, can you see a Marxist concern for the just allocation of spaces? Or a Freudian concern with motivation and impulse? Where and how in the history of the modern movement? Is a milkbottle and a wine bottle inferior to art glass from Finland. If so, why? When a textile is used for curtains, what are its functions? What alternatives are there? If you blow up a letter form and cut it out in hardboard, do you like carrying it under your arm? Try setting verse in Plantin, Univers, Bodoni; other things equal, what difference? Should a table be flat, if so, why? Was Albers right to say that flush joints work in metal and not in wood? How would you make a diagram to compactly suggest the nature of wood as a material? Consider noise, signal, redundance; are there distinctions here that would help you design anything? What makes a Georgian sash window 'work' visually and a Victorian sash gape like a fish? How would you replace either? Have you compared drinking cold water from glass, plastic, china, paper, stoneware, metal? Try listing the requirements you would need to satisfy in designing an eggcup and a teapot. Would you need to know much about the flow of liquids, what makes a bad teapot choke or drip? Have you ever seen a well-designed electric fire? One minute exercise, examine that old designer's friend the Terry angle-poise lamp. Performance aside, what is glaringly inconsistent about it? What is the difference between honesty and sincerity in your work? If you are reorganizing your room, should you 'respect' existing dimensions, if so, why? Are there primary and secondary dimensions? What is the difference between a container and an enclosure, could it be relevant to know? Would you design a prison, if not why not? Is a book a 4-dimensional object, should it be seen with X-ray eyes? Can alertness be summoned and sustained by an act of will? If so, when useful? Would you have preferred this book to be illustrated with comic strips, if so, why? Is it good for people to stretch and reach into inconvenient places. If so, is ergonomics a science? What do you like about your favourite things? Compare pubs closely. What makes a good pub? Location, publican, people who use it, decor, privacy, good beer, comfort ... Also launderettes. What makes one better to be in than another, lighting, surfaces, facilities, efficiency ... how would you make one better? In a cafe consider the tables etc. drawn out as a plan on your drawing board.

Suppose the plan and circulation would read plausibly, the seating comfortable, yet the cafe is thoroughly depressing to be in. What essential decisions might be missing from your plan? (cf. Corbusier, the plan is the generator ...) What is intrinsically wrong about bookcases, or questionable? Is the surface the heart of things? How else will you know them? Is a door handle a piece of information, what else might you prescribe for an outside door? What is a door, irreducibly? Study watch and clock faces. Should a watch be round, square, or? If you were visually aligning with a circle, would you wish to pick up its centre or periphery? (Other things being equal.) Could they be equal?

Appendix III: Extract from the Chairman's report on the National Conference on Art & Design Education at the Round House, London NW1, July 1968

The Conference, which was sponsored by the Movement for Re-thinking Art and Design Education (MORADE), had two objectives:
(1) to promote whatever action was seen to be called for immediately in view of the present situation in art school throughout the country, and
(2) to establish on a national scale the proper bases for further study of art & design education and of the matters that are relevant to it.

Contributors, who were drawn almost equally from the staff and students of art colleges and schools, were invited to speak freely on whatever topics they believed to be important or relevant within the general framework of the following questions: (1) Why art & design education? (2) What is a school of art?, and (3) How should art schools be organized?

The Conference soon found itself to be in agreement that the purpose of art & design education is to develop critical awareness, to allow potentially creative people to develop their aptitudes, to encourage questioning and to stimulate discovery, and to promote creative behaviour. It was also generally agreed that this purpose could not be served except under conditions of freedom far greater than obtain at present – freedom from external control by bodies unsympathetic to and uncomprehending of its purpose, freedom to select students without constraint by irrelevant criteria, freedom to develop courses without regard to inappropriately academic national standards, and freedom from inhibition by too-rigid structures of internal control. The Conference recognized the urgent need for reform by the immediate removal of some impediments but it also recognized that reform in the longer term would need much further study and might well involve the re-orientation of art teaching throughout the educational system as a whole. Voices were not lacking to remind the Conference of the equal need for realism.

A recurrent theme was the relationship of 'Art' to 'Society' and, therefore, of the role or roles – actual and potential – of the artist and designer today. A wide diversity of views was expressed from which it emerged that the need for solidarity in confronting a world unaware of art's value of purposes outweighed the need that might arise for distinguishing differences of function and approach between, say, 'artist' and 'designer'. It was made apparent to the Conference, by the remarks of Sir John Summerson, that even within bodies nominally constituted to represent their views there is an alarming and – in the present situation – possibly crucial lack of fundamental understanding. It was agreed by the Conference, therefore, that a primary function of art education is the extension of understanding and that a world which does not know 'what art is about' will neither be able to use it rightly nor concede to it a proper status. In this 'chicken & egg' situation the need for internal reform is paramount and urgent.

Geoffrey Bocking, Chairman